Dieter Straub

Thermofluiddynamics
of Optimized Rocket Propulsions

Extended Lewis Code Fundamentals

1989

Birkhäuser Verlag
Basel · Boston · Berlin

The work upon which this study is based was commissioned by the Federal Minister for Research and Technology (File: 01 TM 8603- AK/PA 1). The author is alone responsible for the study's contents.

Translated from the German by
C.W. Offermann
Niobestrasse 28b
8000 München 82
West Germany

Author's address
Prof. D. Straub
Institute of Thermodynamics
Univ. Armed Forces Munich
Werner Heisenberg-Weg 39
8014 Neubiberg
West Germany

Library of Congress Cataloging in Publication Data
Straub, Dieter, 1934–
Thermofluiddynamics of optimized rocket propulsions extended Lewis code fundamentals / Dieter Straub: [translated from the German by C.W. Offermann].
p. cm.
Includes index.
Bibliography: p.
1. Rockets (Aeronautics) – Performance. 2. Chemical equilibrium.
I. Title.
TL783.S77   1989
629.47'52 – dc 19

CIP-Titelaufnahme der Deutschen Bibliothek
Straub, Dieter:
Thermofluiddynamics of optimized rocket propulsions: extended Lewis Code fundamentals / Dieter Straub. [Transl. from German by C.W. Offermann]. – Basel; Boston; Berlin: Birkhäuser, 1989

ISBN-13: 978-3-0348-9927-7   e-ISBN-13: 978-3-0348-9150-9
DOI: 10.1007/978-3-0348-9150-9

© 1989 Birkhäuser Verlag, Basel
Softcover reprint of the hardcover 1st edition 1989
Cover Design: Albert Gomm

Dedicated to my wife Brigitte

«Es ist nicht gesagt, daß es besser wird,
wenn es anders wird.
Wenn es aber besser werden soll,
muß es anders werden.»

G.Ch. Lichtenberg

CONTENTS

FIGURES AND TABLES

Part I

Part II

"Dites-moi s'il y a des sectes en géométrie?"
    -Voltaire, L'ingénu-

Preface

This book deals with one of the central themes of thermofluiddynamics
-the process of simultaneous chemical reactions in flows- occurring in
a special limiting case of particular interest for the development of
rocket engines. Although the theory presented here agrees only in certain known limiting states with current textbooks, it is fully compatible with Gibbs' principles. It offers an important contribution to the
understanding of gasdynamic relationships in chemically reacting gas
mixtures. The theory thus opens up new perspectives for the design of
future rockets and air-breathing engines for supersonic vehicles: for
the first time, their development and system integration can be based
on easily comprehensible theoretical fundamentals allowing all relevant
operating parameters to be adjusted to reveal optimal relationships.
This possibility is obviously relevant both for the planning of complex
technical systems and for evaluating engine operation data. It is guaranteed because the theory's validity can be mathematically proven.

This study was stimulated by a NASA workshop meeting and commissioned
by the Bundesministerium für Forschung und Technologie of the Federal
Republic of Germany. Participants met at the workshop in Huntsville,
Alabama, late in February, 1985, to discuss the obvious defects in the
so-called NASA-Lewis Code SP-273 (1971/76). This code, created by S.
Gordon and B.J. McBride, is currently still in use as a standard
throughout the world. Experts at the workshop unanimously recommended
that it be revised and extended.

The study consists of an analysis and correction of the Lewis Code, and
thus supplements NASA's report. The theory presented here was originally constructed for liquid-fuel rocket engines, yet also offers some interesting possibilities for extension to actual RAM and SCRAM jet engines. The numerous comparisons of tested rocket engines in the study

were confined to systems with high-pressure hydrogen-oxygen combustion. This liquid-fuel combination was chosen not only because of its wide-spread current application, but because it also demonstrates how a typical lack of precise information on thermodynamic material properties can have negative consequences in practical applications.

The contractor requested only a brief presentation –Chapter 2, Part II– of the conceptual basis of the 'Munich Method' offered in this study. This basis will be treated in greater detail in an upcoming report.

The consequent use of an equilibrium concept introduced by G. Falk in his Generalized Dynamics is characteristic for the Munich Method; Falk's theory is a consistent extension of the Gibbs thermodynamics also adopted in the Lewis Code. In this study, a state of equilibrium is not defined as "the final state in a self-running process". It is instead characterized by the extremum behaviour of the mixture flow's Gibbs function in the combustion chamber-nozzle configuration. This state can be achieved only under well-defined conditions of free exchange or invariance in relation to the specified Gibbs function variables set according to the chosen process realization. It must be emphasized that such a thermodynamic state of equilibrium is theoretically non-compatible with kinetic models of 'fast' or 'slow' trends to equilibrium: the latter problem "is unsolved today" (Truesdell) and offers no foundation for a valid extension of the Lewis Code!

This study would not have been possible without the persistent interest of W. Finke, the decisive support of the Bavarian State Chancellery and the indispensible assistance of the project coordination team, H.Lösch, M. Weiß, W. Borcherding and G. Precht. K. Gross and S. Gordon have given me considerable encouragement. I would like to thank the experts at MBB, Munich, and my colleagues E. Adams, D. Geropp, and H. Wilhelmi, and particularly R. Waibel, for their advice and help. Some collaborators involved in the project, such as S. Dirmeier, W. Düster, U. Waibel and G. Waldmann, deserve my thanks and recognition for their substantial contributions. I am especially indebted to my friends G. Kappler and V. Lippig and my wife for their continuous and effective support. Finally, I would like to acknowledge the generous cooperation and support I received from C.W. Offermann and B. Zimmermann, delegate of the Birkhäuser Verlag, Basel.

Synopsis

One of the fundamental achievements of science is the ability, under specific circumstances, to assess precisely the qualities of a physical process. Outstanding examples of this ability can be found in classical thermodynamics: Carnot's cycle as well as Kirchhoff's and Planck's laws governing black body radiation provide us with accurate methods of comparison.

These laws are not only practical, but have paramount conceptual significance for engineering as well, since they allow experiments to be related to natural law. Up to now, however, one has not succeeded in finding suitable criteria of evaluation for the physical processes in which gas flows and chemical reactions occur simultaneously. There is a historical and relatively banal reason for this gap in our knowledge: when optimizing a problem in classical (Gibbs) thermodynamics, the mechanics of gas flows usually plays a very minor role as long as the energy conversions resulting from chemical reactions are taken into consideration.

In contrast, flow mechanics today is first and foremost a kinematic theory covering the space-time relationships of physical objects and their transformational behavior. With the exception of some developments in modern gas dynamics, there are few signs of significant work being done on the formulation of evaluation criteria in this field.

This situation is surprisingly true also for linear irreversible thermodynamics and other well-known non-linear field theories. Their purpose is the description of space-time relationships in real processes, but they contribute little to evaluating the complicated and problematic experiments involving such processes. The reason for this is primarily due to the fact that an extremely effective mixture of the newest kinematic notions of rational hydrodynamics and established concepts of classical Gibbs thermodynamics has been concocted - a mixture which invariably leads to dogmatism and sterility. Prominent examples of this type of thinking are the all too familiar terms 'local balance', 'closed and open systems', and 'flow velocity' as kinematic variable, 'pressure' and 'entropy' as 'primitive' properties, 'incompressible fluid', etc.

These conceptions and terms, in a wide variety of forms, are the basis of all modern literature on rocket propulsion. The standard work in this field, by M. Barrère and associates, accurately represents the current status quo of rocket theory in spite of having been written in 1957. This excellent book, which served as one of the foundations of this study, is completely based on the flow tube theory of perfect gases. Although conversion processes are naturally taken into consideration by the authors, their connection with classical gas dynamics is inadequately explained. This interrelationship should be handled in a separate chapter.

The following study presents an ideal comparative process for evaluating real flow processes with simultaneous chemical reactions - a comparative process comparable to Carnot's cycle. The conceptional basis for this so-called "Munich Method" is the Alternative Theory (AT) of thermofluiddynamics, which avoids the dogmatism mentioned above.

The idea of an ideal comparative process was largely influenced by J. Kestin's term 'Lost Available Work'. Both the concepts are methodologically aimed at an absolute optimum realized by reversibility.

The Maxwell-Gouy-Stodola theorem (see BEJAN 1982, p.24) generalized and extended by GLANSDORFF et al. (1955) to monothermal processes of open systems in motion, has also stimulated research. This theorem states that the work of a thermodynamic system is a maximum if the process is completely reversible, the final reaction products are returned to ambient temperature and pressure, and there is a definite expression referring to kinetic and diffusion energy limits. The latter part concerning the flow at exhaust is eliminated by definition in the case of an uniform translation.

The results of the method presented are not only of scientific significance, but also of great practical interest for all current and planned
high performance rocket engines, such as those used for the main engines of the space shuttle, the ARIANE rockets or the future Japanese
H-II system. The results should be particularly important for the engines designed for the super- and hypersonic aircraft being discussed
today (see HÖGENAUER 1986/87). Above all, this study should be considered a contribution made in the spirit expressed by H. Oberth in his
epochal 1929 work 'Paths to Space Travel': "Ideal propulsion is relevant for rocket theory, since it establishes the bench-marks for requirements made on the rocket as well as on the rocket's performance
capacity and the importance of technical improvements." (OBERTH 1986,
p.41)

NASA's assessment methods – in contrast with the "Munich Method" – reveal considerable discrepancies and inacceptable incompatibilities when
compared with test results. These uncertainties seriously question the
achievability of the intended performance improvement. In a September
1984 article entitled "Space Shuttle Not Flexible Enough", the Frank-
furter Allgemeine Zeitung stated that the US Air Force desired large
new rockets specifically because "the main propulsion system and the
fuel pump of the space shuttle still do not attain their expected performance level." Similar difficulties obviously have continued to
exist.
The results presented in this study may be important for the intended
massive performance improvement in the next ARIANE generation, since
that system was also calculated on the basis of the NASA Report SP-273
of March 1976 by GORDON & McBRIDE. The methodology presented in that
report, under the titles 'Equations Describing Chemical Equilibrium'
and 'Rocket Performance', is pragmatically useful but based on questionable physical concepts.

This criticism does not, on the whole, apply to the report's actual
calculations, such as those used for finding the specific vacuum impulse for large ratios between the geometric diameters of a cylindrical
combustion chamber and the nozzle throat. Such results are, considering
the requirements, often sufficiently accurate. The insurmountable problems lie, rather, in the theoretical foundations upon which the report

is based. The report is thus virtually unusable for the design and development of complex propulsion systems: only in individual cases can the user decide if the inherent risks of the NASA-Lewis Code are acceptable. After the Challenger disaster of January 28, 1986, such fateful decisions can scarcely be justified in the future!

Previous discussions with experts from all related disciplines involving the concise presentations of the problems and the intended goals offered in the **first** part of this study have indicated that a presentation of a proposed solution and its perspectives should be given adequate space in a separate **second** part. In addition, investigations of the methodology currently in use or under discussion should serve to illuminate the origin of the above-mentioned incompatibilities. The study critically discusses predominant dogmatisms and their diverse roots. In addition, it illustrates their practical consequences with specific examples.

The new method of calculation is described in detail in the third section of Part II. Its fundamentals are briefly introduced in the second section of Part II.
The presentation is oriented toward calculating data needed in developing rocket engines. Variants on these calculations, suitable for practical applications such as designing the combustion chambers of high-performance plane engines, can be evaluated with the theory presented.

Part II is intended to be, with certain minor limitations, independently comprehensible and applicable: certain repetitions are thus unavoidable. To simplify comparisons of the NASA papers by R.J. Prozan as well as S. Gordon & B.J.McBride, their nomenclature has been adopted in most cases. The last section of the study also contains results of tests, comparative calculations and a pertinent commentary.

Summing up, this study offers a rare example of how an elementary problem of gas dynamics, long since laid ad acta, suddenly assumes great relevance not only in its theoretical field, but in practical engineering applications as well.

"Si un principe cesse d'être fécond, l'expérience, sans
le contredire directement, l'aura cependent condamné."
                                                    -H. Poincaré-

Part One (I): Optimal Comparative Process for Rocket Engines

1. Description of the Problem

1.1 Introduction

In the past few years we have carried out evaluations of gas dynamic
methods currently used in the design of rocket engines (RE). At first
investigations were concentrated on a critical comparison of the physical fundamentals and algorithms used for so-called upper stage engines.
These engines, employed primarily for space missions, are generally
operated by liquid hydrogen (LH) as the fuel and liquid oxygen (LOX) as
the oxidizer. Such rocket engines basically consist of a fuel-oxidizer
feed system, a fuel-oxidizer injection assembly and a cylindrical combustion chamber with adjoining Laval nozzle (see MIELKE 1986, p.92).
Characteristic cross section areas at the end of the nozzle A_E , of the
combustion chamber A_C and of the throat of the nozzle A_T are usually in
the proportions 20:2:1. The general scheme of such a rocket engine is
shown in Figure 1.1.

The rocket engine should produce thrust S_E . This quantity, which ultimately defines every space mission, is closely related to the steady
mass flow rate $\dot{m}$ and the (mass-)specific impulse i_E of the burning gases in the final cross section area A_E. The gases are produced after va-
porization and ignition of the mixed liquid oxidizer and fuel just behind the injector section. **This polynary mixture is the thermofluiddynamic system being investigated here!**

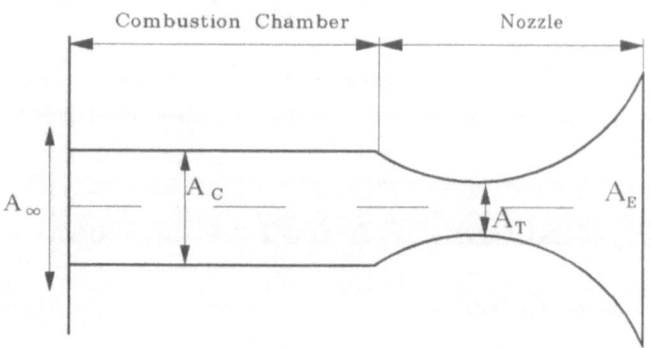

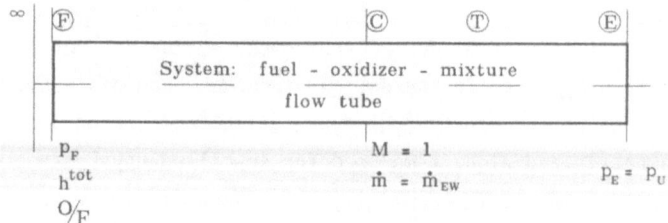

Figure 1.1: The Sequence of States; Assignment to
Combustion Chamber - Laval Nozzle Configuration

The thrust S_E is presently established with an elementary set of definitions that contains all measurable operating parameters. In addition
to $\dot{m}$, the feed pressure p_F and the **nominal** thrust at ambient pressure
$p_U = p_E$, (see MIELKE 1986, p. 439), the set includes:

$$(1.1) \qquad S_E := \dot{m}\, v_E$$

$$(1.2) \qquad S_E := C_F\, A_T\, p_F$$

$$(1.3) \qquad v_E := g_o\, I_{sp,E}$$

$$(1.4) \qquad v_E := C_F\, c^* \qquad .$$

Equation (1.1) was first given by I.W. Mestscherski (1859 - 1935) under simplified circumstances. As a so-called 'rocket equation', it combines two characteristic quantities: the mass flow rate $\dot{m}$ and its specific (second) impulse $I_{sp,E}$. The exhaust velocity v_E of the burning gases corresponds often to the specific impulse $I_{sp,E}$, conventionally coupled to the value of the standard acceleration of gravity g_0 (here 9,81m/s²): one thus has the equation $S_E^\lambda = \dot{m}\ I_{sp,E}$ in mass units (for example, kg).

Equation (1.2) expresses S_E proportional to a force: a term formed from the product of p_F and the nozzle cross section A_T . The proportionality factor C_F usually lies between 1 and 2 (see RUPPE 1982, p. 40): it is a complicated function of numerous influences which, as we know from previous experience, have the effect that the **characteristic velocity** c^* (in Russian literature: 'flow rate complex', see PIRUMOV & ROSLYAKOV 1986, p.204), in contrast to v_E , often may be handled as a problem-specific 'constant' with a given RE geometry. The associated problematics will be specified in Part II.

If one couples the definitions (1.1), (1.2) and (1.4),

$$(1.5) \qquad \dot{m} = \frac{A_T\ p_F}{c^*} \quad ,$$

this relationship – with given values for A_T and p_F – is customarily used to determine the mass flow rate $\dot{m}$ of the burning gases (see RUPPE 1982, p.32); c^* is established from model theories for the RE (see section 3 and SUTTON 1986, p.55), thus introducing the first real physics.[1] The same is also true for the **thrust coefficient** C_F, a factor which allows calculation of the thrust.

One frequently encounters definitions for S_E in technical literature which differ from equation (1.1) (see PENNER 1957, p.158). The expression (see MEBUS 1957, p.16 or BARRèRE et al. 1960, pp.25-27)

$$(1.6) \qquad S_E := \dot{m}\ v_E + (p_E - p_U)\ A_E \quad ,$$

coincides with the above-cited definition as long as pressure p_E in the nozzle exit is equal to ambient pressure p_U (= 'ideal' nozzle).[2] If $p_U = 0$, then S_E corresponds to the **vacuum thrust** (see MIELKE 1986, p. 439). This term is especially relevant for theoretical considerations.

In all practical cases $p_E > p_U$ holds, and a physically transparent relationship between S_E^{vac}/A_E and the pressure ratio p_F/p_E results for some
simple model theories (see Section 3).

Frequently – particularly in older rocket literature and in textbooks
on gas dynamics – the correlation between S_E and v_E is explained in
terms of elementary relationships for a perfect gas. The reciprocal
proportionality $v_E \simeq (\psi_E)^{-\frac{1}{2}}$ between v_E and the square root of the mean
molar mass ψ of the combustion gas in the nozzle exit sometimes leads
one to believe that a reduced value of the mean molar mass would be ad-
vantageous. In an oxygen-hydrogen combustion, the H_2 excess is often
justified in this way (see SÄNGER 1965, p.67). VALIER (1928, pp.130-
131) and later RUPPE (1982, p.44) explain why this argumentation is not
valid for rocket engines.

1.2 NASA Comparative Processes

Our analysis has been based on three methods which were initiated and
sponsered by NASA in a twenty-year period up to around 1970. The FAC-
Method (**Finite Area Combustion**) is a further development of the AFC-Me-
thod (**Adiabatic Flame Combustion**), which, in the version published by
S. Gordon & B.J. McBride (1976), is presently a world-wide standard.
The two proposals made by Prozan – the **Prozan Procedure** (PP) – are not
officially available.

All three procedures are based on current textbook knowledge, taking
into account traditional thermodynamics and reaction kinetics research
up to the early 1960s. One competent reference is S.S. Penner's mono-
graph on propulsion systems, including fuels (see PENNER 1957, p.215).
The original aim of all three procedures is identical:
**the formulation of a calculable comparative process for the steady com-
bustion of fuel-oxidizer mixtures with subsequent ultrasonic expansion.**

This comparative process is useful for evaluating test results independent of items accounting for changes in the real operating conditions,
such as regenerative cooling, mixing entropy just beyond the injector
section, turbulence effects, combustion chamber oscillations, non-equilibrium reactions or shock reflections. **It must be stressed that it is**

**not the primary task of this comparative procedure to simulate such re-
al processes!**

One thus finds all the constituent process conditions always used in
classical gas dynamics, supplemented with 'laws' governing chemically-
reacting mixtures.[3]

Following the suggestion of the responsible technical representative at
the Marshall Space Flight Center in Huntsville, Alabama, the investigations were first directed toward apparent contradictions in the three
methods with regard to the most important **basic properties** (design data
and 'critical' quantities) of a rocket engine. One can get an impression of these discrepancies by examining Table 1. The properties are
determined by Prozan's results (1969, pp.19 & 22) or, for the states $\underline{F}$,
$\underline{C}$ and $\underline{I}$ (see Figure 1), are calculated on the basis of the NASA procedures discussed in Part II. In order to make a comparison with the experimentally obtained thrust S_E , for example, the state properties in
the outlet cross section (based on the given properties in the nozzle
throat cross section) were derived from the known elementary relationships of classical gas dynamics ('isentropic expansion': see OSWATITSCH
1976, pp.92f or Fig.1.2 in Part II). A value $\kappa = 1.144$ for the averaged
isentropic exponent was used for convenience.

Basic Data		AFC	FAC	PP	Experiment
$\dot{m}$	kg/s	247.8	247.8	247.8	247.8
T_C	K	3404	3355	2715	?
S_E	kN	1087	1075	988	1003.4
$I_{sp,E}$	s	447	442	406	412.8

Table 1 : Comparison of NASA-methods by means of J-2
engine data

The properties listed in Table 1 refer to the J-2 rocket engine (see,
for example, RUPPE 1982, pp.69 f and MIELKE 1986, p.213), using known
thermodynamic data for a LH/LOX-mixture with a ratio of the mass rates
of $\dot{m}_{LOX}/\dot{m}_{LH} = 5.5$. It is assumed that the combustion products consist

of the six gas **components**

$$H_2, \ O_2, \ H_2O, \ HO, \ H, \ O \quad .$$

Occasionally the species HO_2 and H_2O_2 are also taken into account, although their thermodynamic data is especially uncertain.

According to the first study by R.J. Prozan, the following properties for the state $\underline{F}$ were chosen (see PROZAN 1969, p.17):

o specific enthalpy of the mixture : $h_F = -225$ cal/g

o injection pressure $\bar{p}_{inj} \approx p_F$: $p_F = 52.62$ atm.

The required thermodynamic data was also taken from Prozan's work: it corresponds to values given in the well-known JANAF tables. The pertinent values for the cross section areas of the cylindrical combustion chamber and Laval nozzle (and their relationships to one another) are:

$$A_c = 0.1722 \ m^2 \quad ; \quad A_c/A_T = 1.58 \quad ; \quad A_E/A_T = 27.5 \quad .$$

Experimentally accessible **design properties** – **feed pressure** p_F , **mass flow rate** $\dot{m}$ and **engine thrust** S_E – for various existing rocket engines are presented in <u>Table 2</u>. All these interpretations, including the values for the respective nozzle efficiency η_D, were taken from G. Schmidt (1986).[*] The data includes engines ranging from the father of all modern high-performance rocket engines, the upper stage Rocketdyne J-2 engine used in the Saturn V rocket, to the impressive values of the Space Shuttle Main Engine (SSME)[4], to the specifications for International Rockwell's small rocket engines (ASE) and the French ARIANE. Rather than simply point out the sizeable differences in performance, which are roughly proportional to the differences in mass flow rates, it is important to note the considerable divergencies in the feed pressure p_F at a concurrently nearly constant ratio of LH/LOX-mass flow rates $\dot{m}_{LOX}/\dot{m}_{LH}$.

[*] from Messerschmidt-Bölkow-Blohm-Corporation, Munich

2)	J2	J2-S	SSME	ASE	HM7-B	HM60-1	
Vacuum Thrust	1001	1171	2090	89.0	59.4	1032	kN
Flowrate $\dot{m}$: Throat	237.5	268.5	471.5	19.18	13.36	237.4	kg/s
Mixture Ratio	5.71	5.85	6.00	6.00	5.30	5.70	-
Injection Pressure	53.1	85.9	207.9	140.0	35.9	103.6	bar
Contraction Ratio $\varepsilon_C = A_C/A_T$	1.58	2.34	2.96	3.66	2.88	2.46	-
Throat Total Pressure p_T	48.7	82.6	202.9	137.9	35.0	100.0	bar
Throat Area A_T	1099	750.4	538.1	31.97	88.50	547.0	cm^2
Throat Temperature	3218	3305	3406	-	-	3267	K
Characteristic Velocity $c^*_{eff} = \dfrac{p_T \cdot A_T}{\dot{m}}$	2253	2308	2315	2299	2318	2304	m/s
C. V. Theoretical c^*_{th}	2326	2328	2329	2323	2349	2327	m/s
Energy Release Efficiency $\eta_* = \dfrac{c^*_{eff}}{c^*_{th}}$	0.969	0.991	0.994	0.990	0.987	0.990	-
Nozzle Outlet $\dot{m}$	241.2	273.6	471.5	19.18	13.51	239.2	kg/s
Nozzle Outlet Pressure (one-dimensional)	0.189	0.181	0.180	-	0.029	0.174	bar
Effective Vacuum Specific Impulse I_{eff}	423.2	436.4	452.0	473.2	448.3	439.8	s
Expansion Ratio $\varepsilon = \dfrac{A_E}{A_T}$	27.5	39.8	77.5	400	82.9	45.0	-
Theoretical Vacuum Specific Impulse I_{th} 1)	445.5	452.7	466.2	485.9	467.2	457.1	s
Thrust Chamber Efficiency $\eta_I = \dfrac{I_{eff}}{I_{th}}$	0.95	0.964	0.97	0.974	0.96	0.962	-
Nozzle Efficiency $\eta_D = \dfrac{\eta_I}{\eta^*_c}$	0.981	0.97	0.975	0.984	0.973	0.972	-

<u>Table 2</u>: Comparison of LH/LOX-Rocket Engines (Data compiled by MBB, Munich)

1) shifting equilibrium according to NASA-SP 273
2) SSME (Space Shuttle Main Engine), ASE (Advanced Space Engine), HM (Hydrogen Moteur)

In this context one can recommend the interesting study of K. STÖCKEL (1985, p.13) on the expected "LH$_2$/LO$_2$ high pressure development", which the author presents as a prospect for the future in his meritorious 'History of the High-Pressure Main-Stream Rocket Engine'. According to him, one can expect combustion chamber pressures up to 500 bar!

For the following argumentation, the astonishingly corresponding values for the respective nozzle efficiency η_I are significant. In practice, the definition

(1.7)
$$\eta_I := I_{sp,E}^{exp} / I_{sp,E}^{theor}$$

$$:= I_{eff}^{vac} / I_{th}^{vac}$$

is used for convenience. Normally the theoretical value I_{NASA}^{vac} is calculated according to the NASA Report SP-273 by GORDON & McBRIDE (1976), and corrected or 'matched' as follows:

(1.8)
$$I_{th}^{vac} := \varsigma \; I_{NASA}^{vac}$$

Used for another calculating method, equation (1.8) is simply a definition for the factor ς. In Table 2, I_{eff} (for vacuum conditions) results from the ratio of the measured thrust to the prescribed mass flow rate at the nozzle end. If one identifies ς with the efficiency of combustion defined by the characteristic velocity c^* (see Table 2), one often finds in literature the expression 'nozzle efficiency' η_D given as $\eta_D :=$ η_I / ς .

Values of $\eta_D \approx 1$ apparently are familiar to every rocket specialist, helping explain why **the methods of calculation used in rocketry generally are considered as experimentally proven** (see PENNER 1957, p.357).

Perhaps such values near one seem more suspect to non-specialists as they are historically obscure with regard to rocket's classic (see NOORDUNG 1929, p.27 f). They result from definition problems with efficiency terms (see BARRèRE et al. 1960, Chapt. 1.5 and pp.340 f). This experience was also the motivation for a critical analysis of the fundamental methods used for calculating the specific impulse I_{theor}.

This analysis is based on the definition given in (1.7). Since the definition is not sufficient for a satisfactory evaluation (as will be shown in Section 4.1 of Part II), the 'thrust quality factor' (equation (4.4) in Part II) is introduced. It is related to a certain extent to two evaluation properties designated as 'discharge correction factor' and 'thrust correction factor' by SUTTON (1986, p.58).

To begin with, a comparison of data was made using the two NASA procedures AFC and FAC together with experimental results. It completely confirmed the experiences from numerous NASA analyses, and certainly prompted additional comprehensive investigations.These were subsequently presented by Prozan: "The adiabatic flame and isentropic expansion solution typified by ZELEZNIK and GORDON ... gave good mass flows but overpredicted thrust and therefore, specific impulse...[The FAC-Method] reduces the thrust to approximately the experimental level but had the disconcerting effect of reducing the mass flow in proportion such that the specific impulse value remained in error." (PROZAN 1969, p.2)

The results of both Prozan studies (and one must note that the second study, completed 12 years after the first, essentially tested the relevance of the first) are in part peculiar. Within the framework of the precision of the experiments and considering the dependence of the calculated specific impulse (in the nozzle exit area) on the choice of the κ-value, the results harmonize neatly with the experiments with the J-2 engine. In addition, they offer encouraging information concerning one of the sorest points in the current Space Shuttle engine (the gas temperature at the point of the narrowest nozzle section): Prozan alleges that one can expect considerably lower values there than predicted by the AFC and FAC methods.[5] This result was quite seductive and made a comprehensive investigation imperative. Such an exploration was also required because Prozan's results now confirmed, in essential details, the experiences of those engineers who believed that theory and experiment 'at these high combustion temperatures' practically are of equivalent quality. This obvious conclusion would mean that all conceivable real influences of turbulent, multidimensional flows with superimposed complicated oscillations, shock layers and simultaneous energy conversions (internal molecular degrees of freedom, chemical reactions, thermal radiation) would neutralize each other in a cooled engine. This being the case, then a relatively simple comparative process could 'simulate reality' for all operational conditions and render the difficult and troublesome study of the real influences utterly superfluous! Is this conclusion in fact correct?

The significance of this study, in the author's opinion, lies primarily in its definitive answer to this question. Naturally it also extensively treats related questions. What would be, in the case of a modern rocket engine, an 'ideal propulsion unit' in H. Oberth's sense of the term? What does an appropriate comparative process 'represent physically', or may such an idealization be expressed in clear algorithms? And how accurately, if at all, do the known procedures – AFC, FAC and PP – describe such comparative processes? To answer such questions, a careful analysis of the theoretical basis of these three procedures is necessary. Only then can a concept be developed for an assertible construction of a comparative process based on physical reality.

> One of the principal objects of theoretical research in any
> department of knowledge is to find the point of view from
> which the subject appears in its greatist simplicity.
>
> —J.W. Gibbs—

2. Definition of an Ideal Comparative Process

2.1 Process Phenomenology in a Rocket Engine

Let's assume that a multicomponent mixture flows within a spatial configuration extending from the fuel injection assembly through the cylindrical combustion chamber and nozzle throat to the nozzle outlet
section. For this thermodynamic system, in which numerous simultaneous
relaxation processes (in the broadest sense) occur, one immediately establishes the following facts:

(1) seen from the standpoint of continuum mechanics, the theory would
    have to deal with a currently insolvable problem if a precise de-
    scription of the extremely complicated space-time behaviour of the
    system along the engine axis is desired. Together with the diffi-
    culty in formulating initial and boundary conditions, even in the
    most optimal case one has to deal with a large set of non-linear
    partial differential equations, primarily of elliptic type; in less
    than optimal situations, with noticeable influences by thermal ra-
    diation, one is faced with integro-differential equations. The ele-
    mentary processes expressed in the individual terms of these dif-
    ferential equations often are very closely interrelated: such in-
    teractions in non-uniform turbulent flames (with their occasional
    dangerous oscillations) cannot be realistically described at the
    present. Estimations are obviously both possible and useful, parti-
    cularly for the delicate heat transfers in nozzle throat regions.
    These may be obtained with boundary layer calculations (see MITRA
    1985 and GEROPP 1987).

(2) Because of the chosen configuration, one can detect many clearly differentiated regions in the system. Within the framework of classical gas dynamics, these regions correspond to the four **configurational states** $\underline{F}$, $\underline{C}$, $\underline{T}$ and $\underline{E}$, as indicated in Figure 1.1 for the respective sections. These states are each marked by specific physical conditions: for a heat-insulated **ideal-typical (total) configuration** consisting of an injection assembly, combustion chamber and Laval nozzle, for example, the states are defined as follows:

$\underline{F}$ is equivalent to a 'state of rest': it is distinguished primarily by a very high value of the system's specific enthalpy h_F compared to the specific kinetic energy $\frac{1}{2}v_F^2$, as well as by high value of the system's pressure, compared to the pressure p_E in the nozzle outlet section.[6] This injection pressure is attributed to the steady mass flow rate from the mixture of oxidizer and fuel in fluid state. It corresponds to the feed pressure p_∞, corrected by the loss of pressure in the injector orifices (see SUTTON 1986 pp.208 f & 223). This irreversible loss depends on injection velocities of the order of about ten metres per second.

There are two types of injectors offering the best performance:

- "Thorough mixing followed by high dispersion of the oxidizer and the fuel

- High dispersion of the oxidizer and fuel followed by thorough mixing" (see BARRèRE et al. 1960, p.387). A fair comparison of the different injector types may be made with the quality factor ξ_b relating the experimental characteristic velocity c^* to the theoretical one. Without considering the heat transfer, efficient injector configurations achieve quality factors near one for gaseous hydrogen-liquid oxygen propellants. Therefore, it is assumed that the difference between p_∞ and p_F is irrelevant for the following theoretical discussion.

$\underline{T}$ relates to the **narrowest cross section** of the total configuration: according to classical flow tube theory, an ideal gas with a p_F/p_E-ratio greater than 2 reaches sonic speed at the throat. In real, multi-dimensional steady flows with chemical reactions in a cooled nozzle, the location of the precise geometric'throat'

usually does not coincide with the position of the system where
the Mach number equals one (see Bray in WEGENER (Ed.) 1970,p.74).

$\underline{E}$ pertains to **supersonic conditions**: the distribution of pressure
in the exit area of the Laval nozzle (see Table 1.5 in Part II)
strongly influences the efficiency of the entire configuration,
particularly with regard to the ambient pressure p_U . The presentation chosen by LIEPMANN & ROSHKO (1957, p.127) for characterizing the influence of p_F/p_U on the nozzle flow offers a good explanation of this phenomenon.

This pressure profile is, in turn, related to the spatial temperature distribution of the steady system in the supersonic part of
the nozzle flow, which itself is highly sensitive to chemical
processes (see Bray in WEGENER (Ed.) 1970, p.152).

$\underline{C}$ indicates the 'state of the combustion chamber': it is primarily
determined by the fact that the (static) combustion chamber pressure p_C usually is notably lower than the 'static pressure' p_F,
due to the 'finite area of the combustion chamber' (see BARRèRE
et al. 1960, pp. 122 f). Apparently the combustion process is
significantly influenced by this pressure drop ('Finite Area Combustion', see PROZAN 1969, p.2 and the literature cited there).

This characteristic pressure drop is one of the major reasons why
many theoretical investigations carried out between approximately
1955 and 1970 on the process of chemical reactions and other relaxation phenomen in Laval nozzles are merely of academic interest.
Almost without exception they apply to the region between $\underline{I}$ and $\underline{E}$,
since supposedly the greatest effects are to be found there. Their
description, however, assumes a complete prior knowledge of the
properties of state in $\underline{I}$, information which one commonly gained by
referring directly to the corresponding, well-known properties in a
'state of rest' (see Bray in WEGENER (Ed.) 1970, p.82).

(3) It is remarkable that – in view of the real physical phenomena –
classical gas dynamics were used to handle situations including relaxation processes (see CLARKE, McCHESNEY 1964/1976). It was noted
that the **Mach number condition M = 1 was also valid in the throat**

of nozzle flows with chemical reactions, if the relaxing flow could
be considered one dimensional, steady and isentropic; that is, having enough small relaxation times, so that the gas mixture is in a
state of "shifting chemical equilibrium" (see BRAY 1970, p.73 and
PETERS 1955, p.226).[7]

It is evident why such a lengthy commentary on seemingly well-known
physical phenomena is necessary. The 'pressure drop' briefly mentioned
above is an exception and has already been sketchily treated by E.
SCHMIDT (1963 p.337). The need for well-founded proofs is apparently
far greater (even in relatively simple cases of thermofluiddynamics)
the moment dissipative processes such as real chemical reactions have
to be taken into account. It is striking that the problems treated in
this study became evident only because earlier authors showed no interest in proving whether the conditions named above for the identity
$M_T = 1$ are even compatible with one another or could at least be physically realized in limiting cases. Even prominent and influential scientists developed theories out of a combination of frictionless balance
equations and dissipative 'rate equations'. Their results are frequently presented without respective commentary and argumentation (see Bray
in WEGENER (Ed.) 1970, pp.81 f), and often end up with 'generalized
rate laws' (p.114) and conclusions which 'naturally' were all 'experimentally verified' (see Bray in WEGENER (Ed.) 1970, pp.147 f). The subject is never discussed with reference to certain discrepancies between
high temperature process thermodynamics and gas dynamical concepts of
mixture flows. One therefore need not wonder that even the elementary
fundamentals of conversion processes hardly received notice in textbooks on modern 'Elements of Gasdynamics'.[8]

2.2 Comparative Process for Relaxing Flows

The objective being investigated here is a thermodynamic system which should pass through an ideal comparative process (ICP), defined as follows:

* the following thermodynamic **assumptions** apply for the process realization:

    (1) the ICP passes through a **sequence of (thermodynamic) states**.
    (2) the ICP is **reversible** and adiabatic.[9]
    (3) the ICP is, in every state, in **thermo-chemical equilibrium**.

* With respect to these assumptions, the following set of **constraints** is valid for the special states $\underline{F}$, $\underline{C}$, $\underline{T}$ and $\underline{E}$:

    (1) 'a state of rest' dominates $\underline{F}$; the pressure and/or the specific enthalpy of the 'mixed' fluid (here the LH-LOX mixture) **before** ignition are given.
    The properties of state in $\underline{F}$ can be determined from the data for the unmixed fuel and oxidizer fluids in state $\underline{\infty}$ (see Appendix 2). Normally, the p_∞-values given in literature on the **injection pressure** correspond to pressure p_F of the unburned fluid mixture immediately behind the injector plate.

    (2) the mass flow density ρv of the system is, in state $\underline{C}$,[10] related to the given duct area A_C by the steady mass flow rate $\dot{m}$.
    (3) the 'Mach number condition' $M_T = 1$ is valid in $\underline{T}$.
    (4) in $\underline{E}$, the pressure (for adjustable states in $\underline{F}$) is at least equal to the known surrounding unperturbed atmospheric pressure p_U (free of schock waves) (see RUPPE 1982, p.62, and BARRèRE et al. 1960, pp.90 f).

* The **system** itself is a **flowing mixture of fluids** that consists, be**fore** ignition (for example, in a J-2 engine), of a binary fuel-oxidizer mixture (here a LH-LOX mixture) with the given mass-flow ratio $\dot{m}_O/\dot{m}_F$ (here: $\dot{m}_{LOX}/\dot{m}_{LH}$), and **after** ignition, of K known gaseous components (for LOX/LH-system: K = 6). The thermodynamic functions of the system are assumed as sufficiently known. This allegation is at present purely illusory for high-pressure-combustion.

* The **ideal-typical configuration** – consisting of a combustion chamber and Laval nozzle for a chosen fuel-oxidizer binary mixture – directly affects the ICP in that the optimal **vacuum**-thrust S_E^{vac} of the configuration (for $p_U \to o$) is partly determined by state $\underline{E}$. The vacuum-thrust definition becomes

(2.1)
$$\boxed{S_E^{vac} := A_E \, p_E + \dot{m} \, v_E}$$

(see equation (1.1) and the endnote pertaining to equation (1.6)); the area A_E of the nozzle exit is, together with A_C , related to the appropriate area A_T through the given 'area ratios'

(2.2) $\epsilon := \dfrac{A_E}{A_T} \; (= 27.5)$ and $\alpha := \dfrac{A_C}{A_T} \; (= 1.58)$

(the values used for the J-2 engine in this study are given in parentheses); A_T is the only choosable geometric parameter in absolute units for the ideal-typical rocket engine configuration. "In practice, the area ratio A_C/A_T ranges between 1.4 and 3."(see BARRèRE et al. 1960, p. 169).

The ideal-typical configuration defines an idealized mapping of the powerplant, i.e. the rocket-motor itself which is assumed to be in steady-state operation.

This somewhat abstract presentation has been chosen on purpose: first of all to avoid misunderstandings by using a compact form; secondly, to make it clear by alienation that the comparative process being investigated here apparently is only roughly or indirectly related to the configuration. This has the advantage that, for this sequence of states (as far as they can be realized), no greater thrust can be achieved than from equations (1.1) or (1.6). Hence, S_E^{opt} is perfectly suited as ideal comparative value for the assessment of corresponding dissipative processes.

Of course one must examine every individual case to see if the formulation of the constraints for the individual states and the linkages (2.1) and (2.2) prevailing between system and configuration are suitable for the specific requirements of real cases (see Section 4).

Such examinations are indispensable because constraints (1) and (4) are not compatible with one another. For the reversibility condition, however, the criterion $p_E \geq p_U$ is essential in order to eliminate the occurrence of dissipative shock waves in the expansion sections of the Laval nozzle. Consequently the pressures p_F and p_U are coupled, assuming the formula $p_E = p_U$ is valid. This line of argumentation offers some interesting practical aspects: one could, for example, create a direct reference to the turbopump capacity of the feed facility for fuel and oxidizer using a given value of p_U from the mission specifications (see KRAMER 1987, pp.188 f).

A theoretical realization of the maximum thrust presupposes, by the way, the coincidence of these two pressures (see BARRèRE et al. 1960, Section 2.4.3.).

For this study, the form of representation chosen completely fulfills the prerequisites and operating conditions used as a basis in the three NASA methods to be analyzed.

It apparently conforms with a theoretically justifiable evaluation procedure for rocket engines; its practical usefulness, however, can be supported only in individual cases.[11] One can even expect an ideal comparative process in a thermodynamic sense if the combustor-nozzle flow is supposed to be isentropic (see E.SCHMIDT 1963, p.315).

The most problematic assumption is the 'chemical equilibrium of the system': it is questionable whether one can ever justify such a definition for flows and formulate the appropriate consistent equilibrium conditions (see VINCENTI & KRUGER 1965, p.252).

The most important assumption is, however, a 'steady one-dimensional flow': that is, the condition for a 'flow tube theory including chemical reactions'. It will be shown that this condition does not imply a superficial drastic simplification of a real, stationary 3-D flow. From the standpoint of thermodynamics, it **provably** offers the optimal possibility of unambiguously evaluating a real – in time and space – physical process through representation in a sequence of states. The process itself will be simulated through the characteristic constraints such as $M_T = 1$ for the corresponding states of the system.

Naturally the authors of the three investigated NASA procedures also dealt with the encountered problems. In Prozan's work, one finds comparatively extensive reflections on the problem, leading to almost metaphysical argumentations for the apparent difficulties and their solutions (see PROZAN 1982, pp.4-5). Yet here, too, one must consider the previously mentioned (more psychological) effect, in which such practiced 'equilibrium computations' seem to correspond accurately enough with empirical experience – especially in view of the fact that "the ensuing hostile environment makes experimental measurements of all but gross quantities virtually impossible" (PROZAN 1982, p.3). The strange remark by R.J.Prozan – "unacceptable errors may still exist in the performance evaluation. This discrepancy has formed a combustion efficiency problem." (PROZAN 1982, p.3) – certainly reinforces the suspicion that the calculation methods of S. Gordon and B.J. McBride and their predecessors more likely were developed as a (rough) approximation of hydrodynamic field equations than as fair algorithms for a proper comparative process. In this same sense they were apparently widely and successfully used. Such practices are highly questionable, particularly in fields such as space technology, where extreme prudence and critical attitudes are generally expected. It must be concluded that the formulation and proof of urgently needed new concepts for 'technical assessments' may be decisively delayed or even prevented by illusory expectations. In this study's section titled 'The Practice with the NASA-Lewis Code', some astonishing indications of such an attitude will be presented.

The AFC method, based on the NASA-Lewis Code, is employed for calculating the **basic data** for the reversible adiabatic flow of an ideal gas in a heat-insulated **Laval nozzle** (compare the standard case in sections 1.2 and 1.5 in Part II). The same is true for the FAC method. Yet in practice, one still finds frequent use of the classical gas dynamic relationships for the isentropic nozzle flow (see Table 1). The flow tube theory is simply modified, assigning the parameter κ (an abbreviation for the ratio of the two specific heats c_p and c_v of the gas mixture in local chemical equilibrium) a suitable averaged value. Possibly justified by practice, this use of an averaged value, however, corresponds theoretically to a polytropic process, and certainly not to isentropic changes. Through such reactions, many nontransparent entropy generation rates occur along the Laval nozzle. The less certain the values at the end of the combustion chamber, the more uncontrollable these rates are. **These properties of state $\underline{C}$ are, however, almost without exception accepted as well-known for the common 'nozzle calculation'. Their determination comprises the major task of the accessible NASA procedures assuming the presumptive coincidence of both the states $\underline{C}$ and $\underline{F}$.**

Another problem, as yet not mentioned, concerns the definition of an **ideal mixture** constituting the 'system'. This ideal mixture-concept is the basis for all previous calculation procedures. Assumptions concerning the fluid state of such a mixture can easily lead to misunderstandings, as experience shows: for high pressure combustion they are questionable (see SCHILLING & FRANCK 1988). For convenience, the relevant relationships should thus be explained in <u>Appendix 2</u>.

"If we are too impressed by the dangers of approximations,
we must dispair at the outset of attaining any rational
understanding of fluid mechanics. However, science was
not built by timid or faltering hands."

-G. Birkhoff-

3. Classical Gas Dynamics of the ICP

In this chapter the ICP will be calculated on the basis of the classical gas dynamics (for this term see L.F.J. Broer in WEGENER (Ed.)
1970, p.3) for a mixture of exchanging elementary ideal gases. Since
the steps of calculation are self-explanatory, the presentation and the
commentary are concentrated on the most significant results. It will be
shown that even in this extremely simple case, all characteristic
features of the ICP, with one exception, are present: in particular the
'pressure drop' in state $\underline{C}$ and Bray's eigen-value problem for the steady mass flow rate $\dot{m}$. The one notable exception is the system's relaxation equilibrium from state to state. This problem has been accurately
described for the first time - without having to accept special "ER-
RORS" - by the Munich Method study (see Part II).

3.1 Relaxing Model Gas as System

The sequence of states $\underline{F}$ to $\underline{E}$ - supplemented with state $\underline{\infty}$ - is shown in
Figure 3.1; the characteristics of each individual state, with the ex-
ception of $\underline{\infty}$, have been described in the previous chapter.

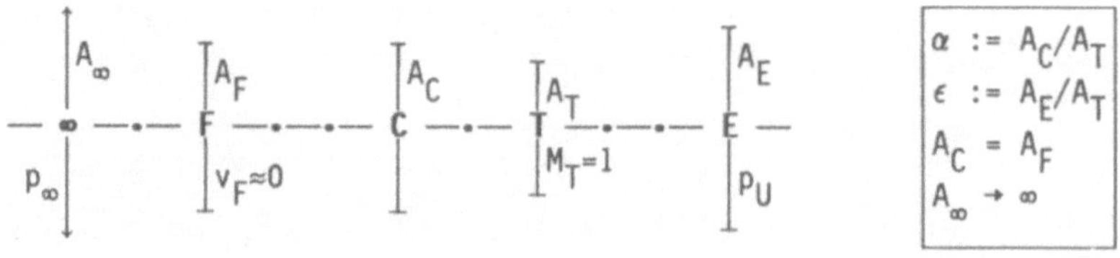

Figure 3.1: Sequence of states of the ICP with given properties

The system should be an **ideal** gas. In order to avoid mathematical complications, the gas should **not** dissociate or, more generally speaking, disintegrate into a chemically-reacting multicomponent mixture, when making the transition from an extremely low temperature at state ⍵ to a higher temperature at state F. This transition is a realistic change of state in a liquid-fuel propulsion system! Such gases do not exist under normal conditions. As a model, however, one can imagine such an ideal molecular gas with a special temperature-dependence of its specific enthalpy h. The molecules' vibrational degrees of freedom between state ⍵ and F may be highly excited and adjust to the corresponding relaxation equilibrium. Diatomic gases such as hydrogen H_2 are especially simple examples.

The model and its thermodynamic foundation are described in Appendix 1; it is defined by the function

(3.1)

$$h := c_p T + \frac{1}{1+(\hat{\gamma})^{-1}} R \theta + e_{10}$$

for the specific enthalpy.

The term c_p indicates the portion of the three translational and two rotational degrees of freedom of each molecule in the total specific heat. In addition, R gives the material-specified gas constant and e_{10} means the ground state energy (not the zero-point energy of the individual molecule!) of the **model gas** at zero temperature. The characteristic material parameter θ denotes the vibrational behaviour of the molecules. For practical calculations, one can identify θ with the characteristic vibrational temperature of diatomic gases given in Table 3:

Gas	H_2	HD	D_2	O_2	F_2	NO	N_2	HCl	CO
θ[K]	6322	5492	4487	2274	1630	2743	3395	4301	3120

Table 3 : Characteristic vibrational temperatures for gases according to SCHÄFER 1960, p.273

Through $\hat{\gamma}$, a relaxation parameter is introduced which is normally dependent on the thermal properties of state, and on the boundary conditions (such as for transfer processes or catalytic influences) or the history of the relaxation process as well.

In the special case of the model gas (index ^), a fair simplification is made: $\hat{\gamma}$ should be independent of the thermodynamic properties of state. In isentropic processes, $\hat{\gamma}$ thus remains unchanged; its numerical value is fixed by the last irreversible change of state experienced by the gas (for example, as the result of a catalytic influence).

Without this simplification, a subsidiary physical condition is necessary for the determination of $\hat{\gamma}$. This situation is typical for equilibrium processes. Their description depends substantially on the specific problems of process realization: in practice, only an approximation of the description is possible. In this case the required isentropic condition can no longer be verified. As one will see, the defect in the NASA-Lewis Code lies precisely in this theoretical limitation. The reasoning for this is rather subtle and thus cannot be treated for the model gas defined above.

In principle the conversion parameter $\hat{\gamma}$ can only be established through measurements. For a model gas, however, one can calculate $\hat{\gamma}$ with equation (3.s) in Appendix 1. For the J-2 engine, the common value of $\kappa = 7/5$ gives $\hat{\gamma} = 0.5783$.

With the temperature derivative of h, one finds c_p for a gas with completely excited rotational degrees of freedom, i.e. a constant. The specific enthalpy h itself is practically reduced to the ground state energy, as long as one uses for T_∞ a very low value $T_\infty \ll T$. This requirement is realistic for diatomic gases which, like H_2 and O_2 used for rocket propellants, are usually found in a liquid or fluid state at extremely low temperatures before ignition.

The model thus appears to be a **perfect gas** which not only complies with the thermal equation of state of an ideal gas, but also has the constant specific heat $c_p = \frac{7}{2} R$. With a gas constant $R = c_p - c_v$ for ideal gases and the **isentropic exponent** $\kappa := c_p/c_v$, the specific heat c_p is expressed through the equation

(3.2) $$c_p = \frac{\kappa}{\kappa-1} R$$

as dependent on R and κ for the model gas the numerical value 7/5 is to be used for κ.

3.2 Flow Tube Theory with Conversion Processes

The classical flow tube theory differentiates between a flow tube with a constant cross-section and one with a cross-sectional area A(x) normally varying in the flow direction where x is the distance along the nozzle center line.

One characteristic feature of classical flow tube equations for **varying cross-sectional area** refers to material systems concerning **steady isentropic flows without internal conversion processes:**

(3.3) $$\frac{d}{dx} (\rho v A) = 0$$

(3.4) $$\frac{d}{dx} [(\rho v^2 + p)A] = p \frac{dA}{dx}$$

(3.5) $$\frac{d}{dx} [h + \frac{1}{2} v^2] = 0$$

this set of equations is usually over-determined.

One can easily show that the combination of equations (3.3) and (3.5) with the isentropic relationship in the form

(3.6) $$\frac{dh}{dx} = \rho^{-1} \frac{dp}{dx}$$

complies identically with the equation of motion (3.4) (see ZIEREP 1976, p.51). This unexpected result is typical for classical gas dynamics: together with the thermal equation of state (3.10) and the caloric equation of state h(T) for the specific enthalpy of a perfect gas (dependent only on temperature T), one needs only four relations for the four variables of state v, T, ρ and p.

The fifth 'redundant equation' is necessary only when conversion processes have to be taken into consideration! This circumstance is elementary for understanding the calculation methods treated in Part II and refers to the inherent difficulties encountered in some of the NASA procedures when the following isentropic relations (3.7) are used for convenience as an alternative to the balance equations (3.4) and (3.5). By using this alternative, one apparently has the possibility of replacing, when needed, an integration of the uncomfortable differential equation (3.4) with an algebraic equation set in the form

$$(3.7) \quad \frac{\rho}{\rho_1} = (\frac{T}{T_1})^{1/(\kappa-1)} \quad ; \quad \frac{p}{p_1} = (\frac{T}{T_1})^{\kappa/(\kappa-1)} \quad ; \quad \frac{p}{p_1} = (\frac{\rho}{\rho_1})^{\kappa} \quad .$$

The index[1] in these equations refers to any chosen reference state with well-known values of mass density ρ, temperature T and pressure p.[12]

If one takes integration constants of the equations (3.3) and (3.5) for the same reference state 1, the algebraic set of equations

$$(3.8) \qquad \rho \, v \, A = \rho_1 v_1 A_1 := \dot{m}_1 = \dot{m}$$

$$(3.9) \qquad h + \frac{1}{2} v^2 = h_1 + \frac{1}{2} v_1^2 \quad ,$$

results, together with equation (3.7). With this equation (3.8), the (steady) **mass flow rate** $\dot{m}$ is defined and the **First Law of thermodynamics** for inviscid flows without heat transfer is formulated (see BAEHR 1966, p.71).

Together with the thermal equation of state of an ideal gas

$$(3.10) \qquad\qquad p = R \, T \, \rho$$

and the caloric equation of state (3.1) for h(T) of the model gas, one can calculate all values of the four properties of state p, T, ρ and v in every state using the equations (3.4), (3.8) and (3.9).

Since the sequence of states which interests us here should occur under certain additional conditions, however, **all reference values are no longer available for every change of state!** Given a supercritical pressure ratio $p_F/p_U > 2$, the Mach number condition $M_T = 1$ results, as known, in the throat area cross section of the Laval nozzle. This generally has important consequences if the feed pressure p_F is given: **the**

mass flow rate ṁ is not a freely choosable value; it must be calculated to be compatible with all other properties of state in the sequence of states. Strangely enough this Bray criterium is universally ignored in practice – most probably because it was not taken into consideration at the time by Gordon & McBride (1969).[13]
As a result of this omission, all affected engines were never designed in view of optimal thermodynamical solutions. This is all the more astonishing when one recalls that this omission primarily affected space projects in which a ratio of payload mass to total rocket mass of about 1:100 was already considered a "natural law", and every additional percent point in favor of payload as a spectacular improvement.

The criticism of a reputed Italian specialist on current design work for turbomachines is an especially appropriate commentary on this particular issue:
"..., the important conclusion is reached that the efficiency of a turbine is set, within a narrow range, simply by the selection of the basic design parameters. All other design phases, though of great importance, cannot yield great efficiency improvements. Similar reasonings hold for other turbomachines."(MACCHI in ÜÇER et al.(Eds.) 1985,p.807).

The setting of the task is equivalent with the representation of all properties of state of the system (= given model gas) under the given set of constraints for the sequence of states $\underline{F}$, $\underline{C}$, $\underline{T}$ and $\underline{E}$ formulated with the aid of algebraic equations. Table 4 is a compilation of the relationships for solving this problem completely. These relationships will be described in greater detail below. The compilation should make clear that, for the known geometric parameters A_T, α and ϵ, the specification of the feed pressure p_F already determines all other values of the steady isentropic flows with conversion processes.[14] This significant situation can be modified especially with regard to the main design data (underlined in Table 4) of a rocket engine: naturally one may use p_U or T_T or $\dot{m}$ or S_E instead of p_F: all other values are adjusted accordingly! The characteristic "ERRORS 1 to 3" accepted for the sake of clarity will be discussed in the following sections.

The algebraic equations are valid not only for the case of classical gas dynamics treated in this chapter, but are also useful for the fundamentals of the Munich Method treated in the second part of the study. In Part II, however, their coherence in concrete mathematical manipulations is far less transparent than expressed in Table 4 as elementary gas dynamic relationships for frictionless conversion processes.

PROPERTY OF STATE	Variables of State			
	F	**C**	**T**	**E**
pressure p	given	motion Eq.	Eq.of state	$= p_U$
density ρ	Eq.of state	Eq.of state	$= \dot{m}/(A_T a_T)$	Eq.of state
temperature T	gas model	fct.of M_C,T_F ("ERROR 1")	energy Eq.	energy Eq.
speed of sound a	gas model	$= v_C/M_C$	$= (\kappa RT_T)^{\frac{1}{2}}$	$= (\kappa RT_E)^{\frac{1}{2}}$
flow velocity v	$\frac{1}{2}v^2 \ll h$	energy Eq.	$= a_T$	$=\dot{m}/(\epsilon A_T \rho_E)$
CONSTRAINTS				
Mach number M	$M_F \ll 1$	fct. of α,κ	$= 1$	$= v/a$
mass flow rate $\dot{m}$	$= \dot{m}_C$	$= \alpha A_T \rho_C v_C$	$= \dot{m}_C$	$= \dot{m}_C$
specific entropy	fct.of T,p,$\hat{\gamma}$	$= s_F$	$\approx s_F$ ("ERROR 2")	$\approx s_F$ ("ERROR 3")
ambient press. p_U thrust S				given definition

Table 4 : Relationships for calculating the properties of state
 - considering the set of constraints; (Eq. = equation,
 fct. = function).

3.3 Sequence of States

* 'Flame' and 'Fluid' States

In state $\underline{F}$ the system is established for the following process by pressure p_F and temperature T_F . Thus, although the flow velocity v_F can be found through the continuity relationship $v_F = \dot{m}_C/(\rho_\infty \alpha A_T)$, the specific kinetic energy $v_F^2/2$ may always be disregarded in comparison to the specific enthalpy h_F .[15] Since the value v_F is completely unimportant for the energy conversion of interest in here, v_F can simply be given the value zero in calculating T_F. Naturally one can modify this characterization of $\underline{F}$ through the substitution $v_F \neq 0$ without any significant complications.

With the energy balance (3.9), the **isobaric** change between the 'fluid' state $\underline{\infty}$ and the 'flame' state $\underline{F}$ is energetically expressed by

$$(3.11.1) \qquad h_\infty + \frac{1}{2} v_\infty^2 = h_F + \frac{1}{2} v_F^2 \qquad ;$$

since in both states the square of the velocity can be neglected with respect to the pertinent specific enthalpy, and temperature T_∞ (considering the physically suitable requirements for a value below the critical temperature of the fuel) may arbitrarily be given the value zero, then the expression

$$(3.11.2) \quad h_\infty = \frac{R\theta}{1+(\hat{\gamma}_\infty)^{-1}} + e_{10} = h_F = \frac{\kappa}{\kappa-1} RT_F + \frac{R\theta}{1+(\hat{\gamma}_F)^{-1}} + e_{10} \quad ,$$

results from equations (3.1) and (3.2). In this equation, the **dissipative catalytic influence** between $\underline{\infty}$ and $\underline{F}$ for the high-energy excitation of the model gas is revealed by both the different values of the conversion parameter $\hat{\gamma}$. For the more qualitatively interesting gas dynamics calculations in this chapter, precise knowledge of the two $\hat{\gamma}$ values is not necessary. In order to simulate a combustion process in a RE by the non-dissociating, vibrationally highly-excited model gas, it is sufficient to replace a common value for the 'flame or combustion temperature' T_F by the characteristic vibrational temperature θ of the corresponding fuel. For example, with the values $\theta = 6322$ K for molecular hydrogen, $\kappa = 1.4$, and $T_F \approx 3000$ K , the result of the difference $[(1 + \hat{\gamma}_\infty^{-1})^{-1} - (1 + \hat{\gamma}_F^{-1})^{-1}]$ is the numerical value 5/3. The following practical formula results in the 'flame temperature' relation

$$(3.12) \qquad T_F := \frac{5}{3} \frac{\kappa-1}{\kappa} \theta$$

for the model gas. With $\kappa = 7/5$, one finds the simple proportionality $T_F \approx \theta/2$ for the characteristic vibrational temperature θ and T_F . It should be emphasized that equation (3.12) is a definition for T_F of the model gas; the inference of equation (3.11.2) was the motivation for the definition! Together with the given feed pressure p_∞ and the physically justified requirement for a nearly isobaric $(p_\infty = p_F)$ 'catalytic' conversion process, state $\underline{F}$ is now established.

For a practical estimation in more complex reactive mixtures, one can also substitute equation (3.12) with a more complicated calculation procedure for T_F. The AFC method is particularly suitable, especially in relation with the Brinkley or RAND algorithms for the calculation of chemical equilibrium properties (see Chapter 1, Part II).

* The 'Combustion Chamber State'

Through a somewhat sophisticated proof (see ZIEREP 1976, p.59), one can show that between ratio α of the two cross section areas of the (cylindrical) combustion chamber (A_C) and the nozzle throat (A_T) on the one side, and the Mach number M_C in state $\underline{C}$ on the other, exists the (nontrivial) relation (see SUTTON 1986, p.38):[16]

$$(3.13) \qquad \alpha^{-1} = M_C \left[(\frac{2}{\kappa+1}) (1 + \frac{\kappa-1}{2} M_C^2) \right]^{-\delta}$$

$$\delta := \frac{\kappa+1}{2(\kappa-1)}$$

The Mach number M_C

$$(3.14) \qquad M := v/a$$

is, by definition, related to the local speed of sound a_C , which, in turn, is based on the definition (compare section 3.3 in Part II)

$$(3.15) \qquad a^2 := (\partial p/\partial \rho)_S$$

and refers to the thermodynamic properties of state in $\underline{C}$.

It is shown in Appendix 1 that, for the conversion processes discussed there, the speed of sound a is in no way identical with the expression $a = (\kappa \, R \, T)^{\frac{1}{2}}$ known from classical gas dynamics. It is, however, a good approximation, above all when one keeps in mind the goal of looking for clarity in the gas dynamics relationships in this chapter. Thus it will be used throughout for the model gas, although it is not compatible with the isentropic relation (3.7.1) – see eqn. (3.s) in Appendix 1 – and causes **"ERROR 1"**.[17]

With the known parameter κ for this type of gas, the relation between $\alpha := A_C / A_T$ and M_C can be iteratively calculated once and for all: in Figure 3.2 this relationship is shown graphically. For the J-2 engine ($\alpha = 1.58$) and $\kappa = 1.4$, M_C is equal to 0.4. This relatively substantial value in comparison with the Mach number (close to) zero in $\underline{\bullet}$ – as the state of fluid model **before** stimulation to high temperature correspond- ing equation (3.12) - is the cause for the previously mentioned 'pres- sure drop' in $\underline{C}$, as opposed to p_F. Using the equations of motion (3.4) for the combustion chamber's constant area section $A_C = \alpha \, A_T$ and the thermal equation of state (3.10), the simple relation

$$(3.16) \qquad \boxed{p_C = p_F \, (1 + \kappa \, M_C^2)^{-1}}$$

results between the combustion chamber pressure p_C and the feed pres- sure $p_F \approx p_\infty$, which may easily be computed with the known κ and M_C pa- rameters (compare with Figure 3.2). With J-2 type engines having rela- tively low α-values, one has to deal with pressure drops in the combus- tion chamber nearly 20% lower than the feed pressure p_F – a result that usually is not taken into consideration.

One should note that the gas velocities in $\underline{F}$ and $\underline{C}$ practically coincide in a combustion chamber with a uniform section. As a result, no pres- sure drop occurs between the 'gas states' $\underline{F}$ and $\underline{C}$. This is not true for the 'fluid state' in $\underline{F}$ **before** the stimulation (=ignition), as it exists in $\underline{\bullet}$. Here the condition $v_\infty^2 \ll v_C^2$ following from $\rho_\infty \gg \rho_F$ for the densi- ties results in a lower pressure p_C in $\underline{C}$.

Using equations (3.13) and (3.16), not only the remaining properties of state T_C (from equation (3.18) given below) and ρ_C (from the thermal equation of state), but also the steady mass flow rate $\dot{m}_C$ can be deter- mined. If $\dot{m}_C = A_C \rho_C v_C$ is established - using equation (3.8) - then v_C

results from the energy equation (3.9) for the states $\underline{F}$ and $\underline{C}$, and ρ_C is expressed through p_C/RT_C .

After a few simple reductions, using equations (3.13), (3.16) and (3.18), one arrives at the expression

(3.17)

$$\dot{m} = A_T \; p_F \; \sqrt{\frac{2}{RT_F}} \; \psi_{max} \; \hat{\Omega}_m(\kappa,\alpha)$$

$$\psi_{max} := \left(\frac{2}{\kappa+1}\right)^{1/(\kappa-1)} \left(\frac{\kappa+1}{\kappa}\right)^{-\frac{1}{2}}$$

$$\hat{\Omega}_m(\kappa,\alpha) := \frac{\left[1 + \frac{\kappa-1}{2} M_C^2\right]^{\kappa/(\kappa-1)}}{1 + \kappa M_C^2}$$

for the steady mass flow rate $\dot{m} = \dot{m}_C$.

The structure of equation (3.17) corresponds to the form common in classical gas dynamics: thus is ψ_{max} the efflux function depending on κ at critical pressure ratio (see SCHMIDT 1963, p.273). The combustion chamber correction $\hat{\Omega}_m(\kappa,\alpha)$ is presented graphically in <u>Figure 3.3</u>: it tends, in the classic 'boiler limit' $\alpha \to \infty$ corresponding to $M_C \to 0$, toward one.

Equation (3.17) has motivated the rocketry community to introduce a property $c.^*$ with the unit of a velocity: if one sets its reciprocal value equal to the product of the square root term and the functions ψ_{max} and Ω_m , a quotient results, consisting only of the experimentally-obtainable properties $\dot{m}$ and p_F , and a given geometric parameter (A_T) (see BARRèRE et al. 1960, pp.61, 98, 181).

The perfect gas isentropic relations (3.7) for changes of state between $\underline{F}$ and $\underline{C}$ were not used in any step necessary for the derivation of equation (3.17). For the model gas with relaxation processes being considered here, the **isentropic relation including conversions**

(3.7.1)

$$(p_C/p_F) = (T_C/T_F)^{\kappa/(\kappa-1)} \; \exp[\theta(1-\hat{\gamma}^{-1})^{-1}(T_C^{-1} - T_F^{-1})] \quad ,$$

is valid (see Appendix 1). The conversion parameter $\hat{\gamma}$ is not subjected to any limitations here, other than the assumption that it should not

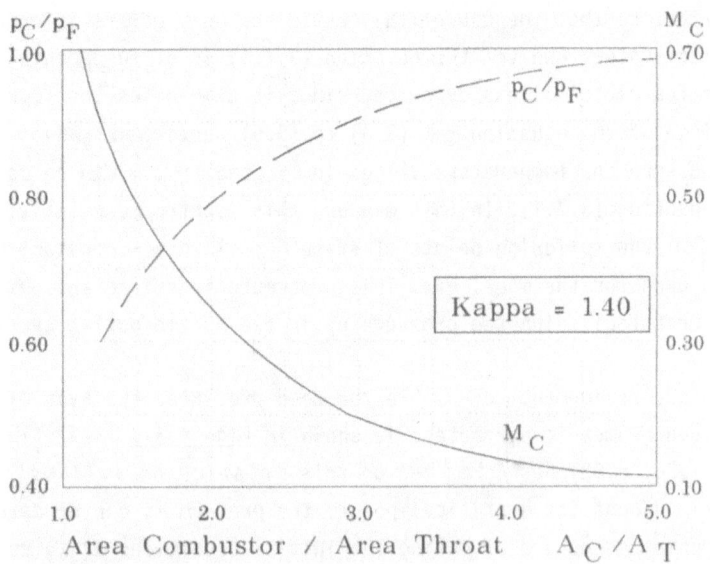

Figure 3.2: Mach Number and Pressure Ratio vs Ratio A_C/A_T

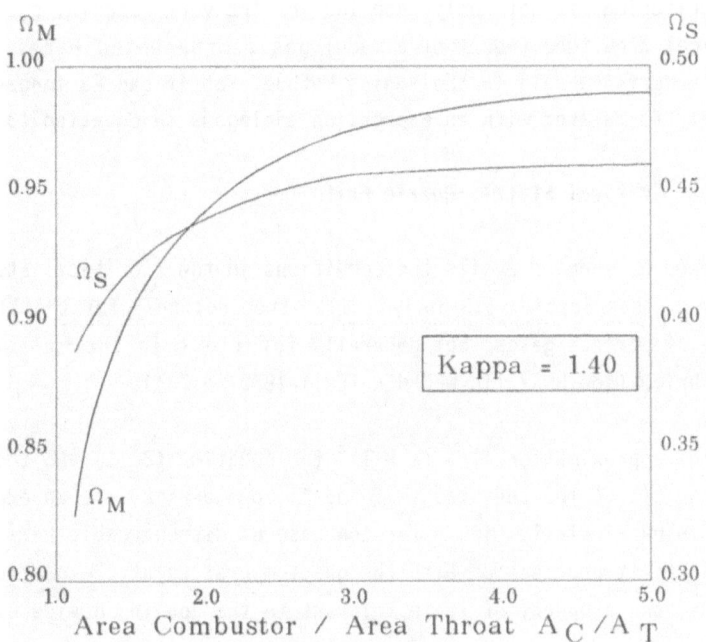

Figure 3.3: Corrections Ω_M and Ω_S vs Ratio A_C/A_T

be dependent solely on the properties of state. It is also not further specified, so that the use of this ratio scarcely offers information of practical value. And yet the equation (3.7.1) is of fundamental signif- icance for classical gas dynamics, since it eliminates the 'over-deter- mination' of the equation set (3.3) to (3.6) mentioned above. Knowing the pressure and temperature values in $\underline{F}$ and $\underline{C}$, $\hat{\gamma}$ can be calculated with equation (3.7.1). In this manner, this isentropic relation can be used for the following points of state[18] , since $\hat{\gamma}$ = constant has been agreed upon for the model gas. This procedure is sufficient, for exam- ple, for establishing the pressure p_E in the nozzle outlet section.

The unique assignment of $\dot{m}$ to the feed pressure, the type of gas and the given geometric parameters is shown in Figure 3.4. Bray (in WEGENER (Ed.) 1970, p.82) has commented on this relation as follows: "If the flow field contains a critical point, the problem is one of determining an eigenvalue $\dot{m}_C$ and of making the initial conditions at x_i compatible with this eigenvalue. Some form of iteration will generally be neces- sary between conditions at x_i and the critical point." Naturally this **Bray criterium** is not only applied in the very special case of the classical flow tube theory of a model gas as presented here! It also plays a decisive role in the Munich Method. Yet it can no longer be ex- plicitly formulated with an expression analogous to equation (3.17).

*   The 'Critical State', Nozzle Exit

The speed of sound prevails the conditions in the 'critical state' $\underline{T}$. This is — see Section 3.3 in Part II — true not only for the flow tube theory of perfect gases, but generally for flows in thermal-chemical equilibrium (see BRAY (in WEGENER (Ed.) 1970, p.73)).

With the approximation $a_T = (\kappa R T)^{\frac{1}{2}}$ for equation (3.15) and the energy equation (3.5) for the states $\underline{F}$ and $\underline{T}$, one arrives at an expression which is particularly simple for the case of disregardable kinetic en- ergy in $\underline{F}$; it provides us with the gas temperature at the nozzle throat section, one property of state relevant to the cooling device of a RE

(3.18)
$$T_T = T_F \frac{2}{\kappa+1}$$
.

General practice confirms that the gas temperature between the states $\underline{F}$ and $\underline{I}$ decreases only slightly and, more importantly, is nearly independent of other variables![19]

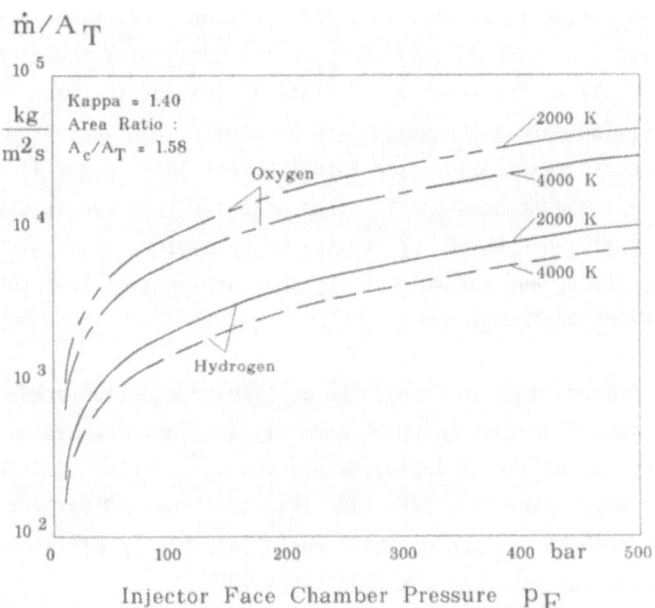

Figure 3.4: Mass Flow Density dependent on p_F

Since all properties of state in $\underline{C}$ are known, all state variables in $\underline{I}$ may be established, through the mass flow rate $\dot{m}$, the thermal equation of state, the energy equation between $\underline{F}$ and $\underline{I}$, as well as the Mach number condition $M_T = 1$. For verifying the isentropic relation one can use equation (3.7.1) between $\underline{C}$ and $\underline{I}$, without having to know .the course of the nozzle area cross section along the axis.[20]

The pressure calculated with the equations of state (3.10) notably does not coincide with the value obtained from equation (3.7.1), using T_T from equation (3.18). The discrepancy is a result of "ERROR 1" (see Table 4). Naturally an agreement can be produced by varying the data in $\underline{C}$ and by using an equation for T_T corresponding to equation (3.18) but not containing "ERROR 1". In such a case, however, equations (3.16) and (3.17) are degraded to physically noncommittal iteration aids and the theory loses its transparency. Thus it is acceptable to dispense with a verification of the isentropy and tolerate **"ERROR 2"**.

The properties of state at the nozzle exit are fully determined by the energy equation (3.9) for T_E , the isentropic relation affecting the conversion processes (3.7.1) for p_E , the continuity equation (3.8) for v_E , as well as the equation of state (3.10) for the mass density ρ_E (Case 1). The more interesting Case 2 results, however, from the demand for a process realization (see FALK & RUPPEL 1976, p.248 f) quantified by $p_E = p_U$. The temperature T_E, then, results from the isentropic relation (3.7.1), for which the conversion parameter $\hat{\gamma}$ is already known. Mass density ρ_E and gas velocity v_E thus are derived from the equations of state and of energy.

It is evident that in Case 2 the continuity equation produces a value for the mass flow rate $\dot{m}_E$ which normally does not coincide with $\dot{m}_C$. The iterative correction is appropriately made by varying the feed pressure p_F until both mass flow rates are identical. One obtains a functional relation between p_U and p_F which could possibly be of interest for the construction of the engine's feed aggregates.

Case 1 is only significant if "ERRORS 1 and 2" have previously been eliminated iteratively in the same manner. In Case 2 there is the additional possibility of assuming the ambient pressure p_U instead of p_F as a mission-conditioned parameter.

In order to retain the physical significance of the algebraic equations derived in the initial sections of this study – for example, $\dot{m}$ as a resultant operational parameter and not as a freely selectable value – Case 3 is required: dispensing with a verification of the isentropic requirement made by equation (3.7.1) and limiting p_E through the condition $p_E = p_U$ (except in the case of a vacuum impulse). Under these circumstances, **"ERROR 3"** is the only possible choice!

Of particular interest is the calculation of the thrust[21] using the properties of state in $\underline{E}$. According to the definition (1.6), equation

(3.19)

$$S_E = \epsilon^{-1} A_T p_F \left[\frac{p_F}{p_U}\right]^{1/\kappa} \hat{\Omega}_S(\kappa,\alpha)$$

$$\hat{\Omega}_S(\kappa,\alpha) := \kappa \left[\frac{2}{\kappa+1}\right]^{(\kappa+1)/(\kappa-1)} \frac{\left[1+\frac{\kappa-1}{2} M_C^2\right]^{\kappa/(\kappa-1)}}{1+\kappa M_C^2}$$

results for $p_E = p_U$.

For the vacuum condition $p_U \to 0$, an indefinite expression results, since the condition $p_E^{\frac{s}{}} = p_U$ in this special case can only be fulfilled with an infinitely large exit area ratio ϵ.

Figure 3.5 shows the curve of the vacuum thrust S_E^{vac} according to equation (2.1) depending on p_F/p_E . It is immediately obvious that the function $\hat{S}_E(p_F/p_E)$ has a minimum! As one can see in this plot, S_E and S_E^{vac} correspond asymptotically with the large (p_F/p_E) values.

Figure 3.6 presents, for the model gas $(\kappa = 1.4)$, the curve of thrust S_E/A_T related to the unit area and depending on the pressure ratio p_F/p_U , with the feed pressure p_F and the exit area ratio ϵ as parameter for the J-2 engine $(\alpha = 1.58)$. Besides values for A_T, α, ϵ, κ, p_F and p_U, no additional information is necessary for calculating S_E.

* **Characteristic Parameters, Summary**

It remains to determine both of the parameters c^* and C_F introduced in Chapter 1, according to the equations (1.5) and (1.2). Both may be easily calculated on the basis of the results presented in the sections above. The following formulas result (see BARRèRE et al. 1960, pp.97f):

    - for the **characteristic velocity**

(3.20)
$$c^* = \left[\sqrt{\frac{2}{RT_F}} \; \psi_{max} \; \hat{\Omega}_m(\kappa,\alpha)\right]^{-1}$$

– for the **thrust coefficient**

$$(3.21) \qquad C_F = \epsilon^{-1} \left[\frac{p_F}{p_U}\right]^{1/\kappa} \hat{\Omega}_S(\kappa, \alpha)$$

– for the **specific (second-) impulse**

$$(3.22) \qquad I_{sp,E} = (\epsilon\, g_0)^{-1} \left[\frac{2}{\kappa+1}\right]^{\kappa/\kappa-1} \sqrt{\frac{\kappa+1}{2} R\, T_F} \left[\frac{p_F}{p_U}\right]^{1/\kappa}$$

according to the equations (1.3) and (1.4).

Parameter c^* is dependent only on α and T_F in examples of the gas model more precisely specified (θ variable, $\kappa = 7/5$) by equation (3.12) and thus is a constant with a given geometry. Parameter C_F as well as $I_{sp,E}$ are dependent on the relevant process parameter p_F/p_U.

With equation (3.22), all relations – which are derived from the given and typical conditions for classical gas dynamics governing the theory of a (nearly) ideal comparison process, and which can be expressed algebraically – are compiled. **This theory takes conversion processes into consideration, but is valid only for a purely artificial gas model.**

Yet the results indicate a number of essential features which one also finds theoretically and experimentally confirmed in real rocket engines. The insensitivity of the characteristic velocity c^* to the operating parameters, for example, is apparent. In addition, the relatively low chance of affecting the gas temperature in the nozzle with geometric or physical properties is evident. Especially important and practically relevant are two physical phenomena which, when ignored as in the NASA calculation methods, can cause serious problems. These two items already arise in the classical flow tube theory previously presented: the 'pressure drop' in a combustion chamber with a finite section, and the unambiguous establishment of the steady mass flow rate $\dot{m}$ as an eigenvalue. Only the important dependence of the gas temperature on the mass flow ratio cannot be simulated by the simplified model.

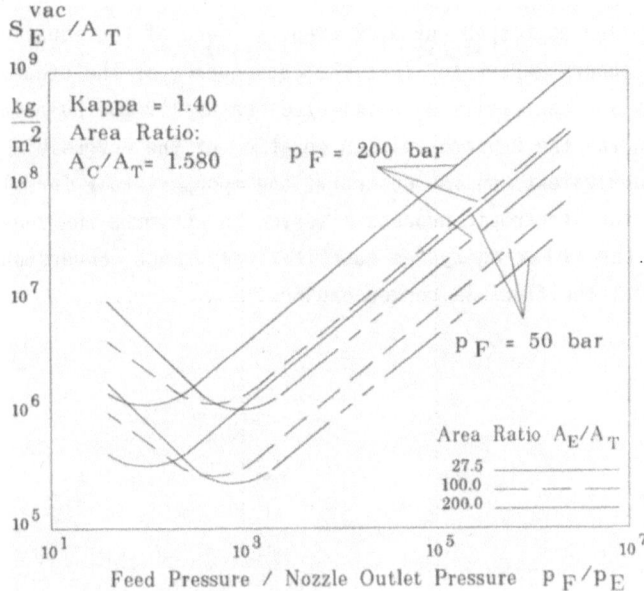

Figure 3.5: Thrust per Square Unit vs Pressure Ratio

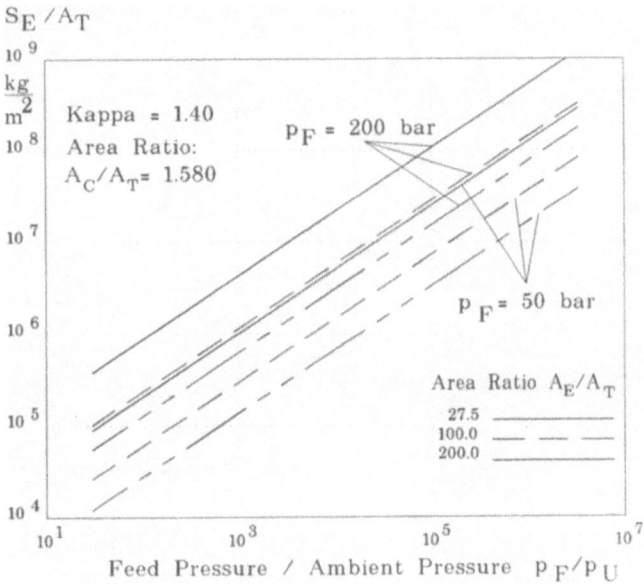

Figure 3.6: Thrust per Square Unit vs Pressure Ratio

More of theoretical interest is the perception that the widely-held preconception expressed by J. Zierep - "one of the equations (3.3) - (3.6) is supernumerary and thus the consequence of the others" (ZIEREP 1976, p.51) - can easily be eliminated. Naturally the Eulerian equation of motion and the Laplace-Poisson equation of the reversible adiabatics are not equivalent: on the contrary, the appropriately formulated relationship for isentropic processes serves to estimate the conversion degree of the relaxing system quantitatively. Such conversion processes are typical for flows in rocket engines.

"Rationality is a quality of treatment, not of the subject treated."

-C. Truesdell-

4. Design Criteria for Rocket Engines

The ICP proposed in this study and defined by a specific sequence of states relates to a thermofluiddynamic system. If one uses this comparative process for calculating an idealized model gas and uses terms which are characteristic for rocket engines (RE) - thrust, nozzle throat, injection or feed pressure, etc. - these terms, although considered commonly as convention, must in fact be fixed in their meaning within the sequence of states through mathematical relations to establish their precise significance in the ICP. It becomes immediately clear that the measured thrust of a rocket engine, S_E, cannot simply be attributed to the corresponding and also experimentally-determined values of the mass flow rate and the velocity of the burned gases in the nozzle outlet cross section (A_E), using the definition $S_E = \dot{m} \, v_E$. One has to ask which value of v_E should be used here, when one assumes that a geometric multidimensional, perhaps turbulent velocity field actually occurs in A_E . Do the technical measuring procedures assure (at least in principle) that the respective mean value of the velocity profiles coincides precisely enough with the value $v_E := S_E/\dot{m}_E$ - a value which, in turn, should be linked with the pressure ratio p_F/p_U, determined by the given feed pressure and registered ambient pressure? Or: which relationship exists between the pressure p_E of the flow tube theory and the often considerably different pressure on the wall of the nozzle in the outlet section? (See the values for various rocket engines in Table 1.5 of Part II.)

Alone these few, practice-oriented examples offer sufficient evidence that **the ICP theory may not be confused with the theory of a RE. Since the ICP is not a physical model for rocket engines it has its limitations, yet also a great flexibility for practical applications.**

Under these circumstances, an obvious question arises: **to what extent and with what reliability can the ICP be used for designing and evaluating a rocket engine?**

This question is by no means trivial, and its usual deliberate disregard in rocket programs inevitably causes costly mistakes and annoying misunderstandings. This problem will be treated in detail in Part II. In this chapter the answer to the question will be limited to two aspects:

□ from the point of view investigated by E. Macchi in his wise study: "Design Limits, Basic Parameter Selection and Organization Methods in Turbomachinery Design",

□ from the point of view discussed by U.G. Pirumov & G.S. Roslyakov in their work: "Main Principles of Choosing a Jet Engine Nozzle".

Both works were recently published and are of great importance.

In two new papers NASA scientists emphasize the need for design criteria, design procedures, and design philosophy (COOPER & SCHEER 1988 and DAVIS & DIXON 1988). The latter authors point out that, "conceptual design codes must remain simple enough that many structural configurations can be explored." (p.42).

A first instructive example has been offered by the study on air-augmented rockets based on a small N_2H_4/N_2O_4 motor. Author D. Rickeard (1973) has defined a simple model to calculate the maximum thrust in steady cruise flight.

4.1 Design Procedure

The notable investigations made by E. Macchi have already been referred to in Section 3.2 above. His study is of particular interest because there are astonishingly few works devoted to the problematics of development associated with the design of highly expensive machinery. Macchi's exposition is so precisely formulated that one need only quote his essential thesis and most important conclusions:

"The preliminary definition of basic turbomachinary parameters can be obtained: 1) by an independent procedure, which will be referred to in the following as "optimization", or 2) by a development of existing units by some of the procedures indicated (scaling, similarity, etc). The various aspects related to the first procedure, i.e. the selection of the basic design parameters of a "new" machine, will be the subject of this lecture. The analysis will be limited to the one-dimensional (or mean line) approach which is just the very first step in the design of a turbomachine. However, the importance of this phase should not be underestimated: for instance, it is well known that the efficiency of a properly designed axial-flow turbine can be predicted with fair accuracy (say, 1 or 2%) by the adoption of simple mean line analysis methods which incorporate proper loss and flow angle correlations ... If, as it is the author's opinion, the above statement is correct, ...[22]

Obviously, the above reasoning was not made to underestimate the importance of the role played by very accurate blade design procedures or flow analysis made nowadays possible by modern computers, but to stress that the first assumptions on turbomachine main characteristics, which are input data to subsequent calculations, are of great importance, too. Hence, the adoption of sophisticated methods for their selection are justified.

The technical literature is surprisingly poor of indications on the selection procedure of the main design data of turbomachines. Possible explanations to this situation could be the following: most industrial machines are designed as development of previous units, and great part of design parameters are neither selected nor optimized, but simply assumed equal to the ones of existing models; "conventional" design parameters, based on experience, can be selected in many fields of application; last but not least, the subject looks relatively elementary, and therefore not very attractive for the researchers. Whichever the reason, the popularity of the few papers dealing on the subject (for instance, (2), (3) and (4))[23] seems to demonstrate the need of reliable methods for the selection of design variables.

2. DEFINITION OF THE PROBLEM

For simplicity, explicit reference will be made in this section to the case of axial-flow turbine design. However, most reasonings hold for all other turbomachines.

2.1 Input Data

a. Thermodynamic properties of the working fluid ...
b. Thermodynamic conditions at inlet ... and outlet ...
c. Flow rate or power ...

2.2 Design Procedure

The design procedure can be looked at as an optimization problem. As in all optimization problems, four main points should be made clear: (1) definition of the function to be optimized. (2) Independent variables; (3) constraints which limit the range of the search; and (4) relations between function, input data and variables ...

A design procedure must be established which allows: 1) the calculation of the target function for assumed values of the optimizing variables and 2) the control that the obtained solution is within the constraints discussed..."(Macchi in ÜÇER et al.(Eds) 1985, pp.805 f).

Macchi's recommendations are obviously valid for engines used in the aerospace industry. In this field of technology, a mission-related optimization of all components in an engine should virtually be taken for granted. In such complex projects, the decisions to 'optimize' any particular framework plan are naturally often vehemently debated, since such orientations are generally more qualitative in nature. One need not exaggerate like rocket pioneer R. Nebel, who equated rocket technology with "the standard for the world-wide reputation of a civilized nation today" (NEBEL 1932, pp.4-5) and called for "new high-tech performances" under the motto "We want to fly higher, we want to fly farther; grant us this flight." Yet a transparent, politically attainable, scientifically acceptable and quantitatively definable project is only possible when it is based on solid fundamentals of a few elementary design aspects. H.O. RUPPE (1985) has provided an exemplary illustration. In space technology and some other areas of so-called high-tech, the design optimization of a concrete propulsion project is usually undertaken with the help of the NASA-Lewis Code. Using such optimization

procedures D. Manski and J.A. Martin (1988) recently have given a concise example for the evaluation of innovative rocket engines. Yet as
one will see, none of the relevant NASA performance programs is adequate for meeting these critical requirements.

In contrast, the Ideal Comparative Process (ICP) is an instrumentarium
in Macchi's sense for the construction of a **target function** as the prerequisite for the optimization of a rocket engine. Insofar as the energy conversion of the burned fuels inside the various parts of a rocket
engine can be represented by the sequence of states in the ICP, one has
the possibility of systematically studying a representation of the
optimal relations between the most effective variables of the ICP. The
proof of optimization is – as can be seen in Chapter 3 – relatively
easy to produce on this theoretical level.

Naturally such a target function does not consist of only one single
equation, but rather a conglomerate of mathematical relations. With
certain constraints, they allow clear identification of the goal one is
looking for: connections between the primary design properties of a
rocket engine. Ultimately it is the engineer's responsibility to decide
just how far he can apply the results of this target function in the
design of 'his' particular RE: the engine is, after all, only one part
of a highly complex carrier system, its cycle performance is related to
the engine mass and to some trajectory and vehicle sizing models (see
MANSKI & MARTIN 1988, pp.3 f).

In this context a comment on the exclusively **one-dimensional** flow of
combustion gases assumed for design calculations may be useful. This
type of flow is familiar from the flow tube theory. As S.A. Shcherbakov
has demonstrated, the rotational-symmetric nature of a steady nozzle
flow influences the local formation of field variables so strongly that
the operating properties obtained through their integration, such as
thrust, diverge from the pertinent flow tube results. For a perfect,
dissipation-free gas ($\kappa = 1.4$), he has drawn the following conclusions:

(i) "the specific thrust in the supercritical regimes exceeds the
value determined using the one-dimensional theory for the same
stagnation parameters of the gas;

(ii) in the subcritical regimes the specific thrust of the contracting nozzle is equal to the value found from the one-dimensional
approximation." (SHCHERBAKOV 1983, p.992).

Judging by these conclusions, it can be assumed that, when using the flow tube theory, one is on the 'safe side': every real nozzle reactive flow can deliver an theoretically attainable, optimal thrust which is not less than that obtained in the one-dimensional ideal comparative process. (See the comments by PIRUMOV & ROSLYAKOV 1986, pp.636 f, on the influence of 3-D effects in nozzles.)

4.2 Influences of Real Flows

The theoretical abilities mentioned in the previous section may also be utilized with considerable advantage, within certain limitations, in the planning and execution of complex experimental programs. Two possibilities may be considered:

(1) Variations of the functional interdependence of the primary engine data, such as feed pressure and thrust of a RE, whereas the optional experimental parameters, such as the steady mass flow rate, are chosen according to the ICP.

(2) The assessment of a RE through appropriately-defined standards of evaluation.

For the selection of such standards of evaluation there are numerous suggestions offered in technical literature. They are naturally tailored to specific projects in which engines are to be employed. Thus, for example, the projected HM 60 engine is, "for reasons of reducing costs and risks in relation to other hydrogen/oxygen engines, to be kept at a low technological level." (See MANSKI 1986, p.51). Regardless of the particular objective to which such ambiguous standards are applied, they ultimately lead to more or less concisely formulated standards of evaluation.

There is not adequate space in this study for a more detailed analysis of the problems relevant to (2) above. Commendable advice for practical applications is offered, for example, in a comparitive study of four different ramjet combustion efficiencies, carried out in 1984 by W.J. Bergmann. (See SUTTON 1986, p.56.)

Of particular interest in this context are the extensive investigations
done by D. Manski. In a recently published work (MANSKI 1986), he has
evaluated the influence of various engine cycle parameters on the payload mass of carriers with differing configurations. His numerical
analysis of performance, structure and motor mass as well as flight
trajectory – based on models – offers a convincing demonstration of
the extremely complex problem of defining an optimization process for a
complete carrier system.

Of major significance for this study is his observation that **all these
standards of evaluation are based on a thermodynamic reference process**:
"the performance model is split into two parts. The first part consists
of a gigantic data file, in which precalculated thermodynamic properties are stored as a multidimensional function ... The second part of
the performance model takes into account the effect caused by different
engine cycles." (MANSKI 1986, p.55). As a rule, the NASA-Lewis Code of
S. Gordon & B.J. McBride is used to describe this reference process
(see BERGMANN 1984, p.130).

D. Manski certainly is not exaggerating when he tersely notes that the
NASA-Lewis Code for the calculation of the thermodynamic performance
data has assumed a **normative** character in rocket technology. (See MANS-
KI 1984, p.114.)

At this point it is important to note that the quality of this process
has varying degrees of influence on all conclusions derived from it
– for example, the 'optimal operating parameters' of a RE. Considering
the importance of the reference process discussed in the NASA-Lewis
Code one would assume that it would be officially questioned because of
its serious theoretical errors. In fact, quite the contrary has long
been the case: in industrial practice, the code is utilized for the
approximate calculation of thermodynamic properties of state of the
burned gases along their flow path from their injection into the combustion chamber to the nozzle throat. Temperatures in the nozzle throat
cross section determined with this method are then correlated with
experimentally-obtained values of feed pressure, mass flow rate and
thrust. "The modus operandi is thus based on the reliability of a calculation procedure for complex chemical equilibrium compositions."
(BERGMANN 1984, p.130).

Except for the numerical data often practicable in individual cases, one of the motives for using such methodologically questionable 'matching procedures` (which frequently produce 'efficiencies` better than one), may be a physical phenomenon expressed in older rocket literature. In 'classical` manuals for the "calculation of rocket engines" (MEBUS 1957, p.18), there is the so-called 'inner efficiency in a perfect combustion`. This is defined (see SÄNGER 1933, p.40) as

$$(4.1) \qquad \eta_i := \left[\frac{v_E}{v_{E,max}}\right]^2 = \frac{T_F - T_E}{T_F}$$

and expresses the ratio of the average gas velocity in the nozzle exit to its fictive maximum value (or its square!) and then – for an equivalent perfect gas – an isentropic change of state from injection into the combustion chamber to the nozzle outlet. Expression (4.1) is formally identical with Carnot's efficiency of a heat engine. Its most important characteristic lies in the fact that its theoretical optimum coincides with its maximum possible limit only by means of a heat sink, which is maintained on the temperature level T_E negligible against the flame temperature T_F.

Since LH/LOX rocket engines even before the Second World War produced η_i-values up to 0.803 – the measured values were first given by H. Oberth and were subsequently quoted by E. Sänger – it is not surprising that **the classical flow tube theory and its modifications for burned gases have been considered virtually to the present day as a sufficiently accurate description of the complex physical processes in a rocket engine.**
Without doubt, such use of the ICP is – above and beyond its original application as a **planning and evaluation instrument** – both physically ingenious and practically justified, as long as the reservations mentioned are not disregarded and thus distort an appraisal of the real conditions in an engine. One (translated) sentence from the preface to Eugen Sänger's famous work 'Raketenflugtechnik` is as valid today as it was when he wrote it: "Since rocket flight is, above all, a technical problem, the book appeals first to engineers and their way of thinking." Thus it is perhaps appropriate here to refer to a particularly useful book with some practical formulas, recently published by U.G. PIRUMOV & ROSLYAKOV (1986, pp.173 f).

It is evident that an expensive program of experiments is unavoidable
for optimizing a rocket engine and providing a more precise analysis of
the causes for all occurrences of thrust loss. Since this is not part
of this study, only a few remarks are made here on the subject. The degree of difficulty of such rocket engine optimization must be obvious
when one bears in mind the efforts which have to be made to coordinate
the measured bulk thrust- and mass flow-values with the flow field of
the burned gases. The expressions for the steady mass flow rate $\dot{m}$ and
for the momentum I in the direction of the nozzle axis,

$$(4.2) \qquad \dot{m} := \int_{A_E} \rho(\mathbf{v} \cdot \mathbf{n}) \; dA$$

$$(4.3) \qquad I := \int_{A_E} (\rho[\mathbf{v}\mathbf{v} \cdot \mathbf{n}] + p\mathbf{n}) \cdot \mathbf{e} \; dA$$

clearly indicate the complexity of this effort: in equation (4.3), for
example, the dyadic product **vv** of the local flow velocity **v** must be
multiplied scalarly with the normal vector **n** of the area element dA;
the result subsequently added with the local pressure p in normal direction, multiplied with the unit vector **e** in the axis direction, finally have to be integrated.

When computing equation (4.3), one must assume that velocity **v** is not,
as a rule, parallel to the nozzle axis, and the pressure p as well
as the Mach number vary considerably over the cross section. This is
caused by various forms of dissipation in the gas. The three principle
forms are (1) fluiddynamical, (2) those associated with the cooling of
the RE and (3) the forms coupled with the diffusion and relaxation processes in the burned gas. In 1952 S.S. Penner offered a transparent
schematic representation of the numerous relations between the complex
physico-chemical processes during stable combustion in rocket engines
(see PENNER 1957, p.359).

Under the fluiddynamical irreversibilities, not only do the viscous
boundary layers appear, but flow instabilities and compression shocks
in divergent parts of the nozzle as well. GEROPP (1987) has published
an excellent study for channel or nozzle flows of non-uniform gases
without conversion processes. This allows all significant effects to be
estimated with algebraic equations, and also takes laminar and turbulent boundary layers into account.

It is unclear to what extent one can speak of turbulence effects in the burned gas having relatively high temperatures and low density. The reference Reynolds number for the laminar-turbulent transition of a flow field is clearly established only for isothermal, practically incompressible flows. For relaxing compressible fluids, additional indicators such as the Mach number and the Damköhler number are significant. Therefore, Reynolds numbers related to nozzle throat conditions (WURST 1987, p.12) are of the order of 10^7 for all rocket engines – with data compiled in Table 2 – and may cause semantic misunderstandings if flow instabilities are simply associated with turbulence phenomena in classical hydrodynamical terms.

Such problems are also particularly difficult to deal with theoretically, because even complex numerical solutions – such as the so-called 'Slender Channel Approximation' of the Navier-Stokes equations – can only be carried out with constant κ-values when estimating the real effects in nozzle flows (see FIEBIG & MITRA 1988). If energetically dominating reaction kinetics are disregarded, then serious studies concerning the stability of flow forms in hot burned gases are impossible.

Even the simplest theories of heated or cooled tube or nozzle flows (see OSWATITSCH 1976, pp.136 f) have already clearly indicated the possibly considerable influences of the convective heat transfer (see STOLL 1987) or radiation exchange with the surroundings on the local thermofluiddynamic state of the burned gases. One can no longer expect, for example, a Mach one line in the nozzle throat cross section. D. Geropp's study mentioned above allows a realistic estimation of all these influences.

This consequence is significant for the physico-chemical conversion processes, in particular for the relaxation processes affecting the diverse vibrational degrees of freedom (see CLARKE & McCHESNEY 1976, pp.252 f and VINCENTI & KRUGER 1967, pp.197 f). They are often more or less frozen in the expansion section of the nozzle, and their calculation generally requires knowledge of the thermodynamic conditions in the nozzle throat cross section. For such calculations (see Bray in WEGENER (Ed.) 1970, pp.90 f and PIRUMOV & ROSLYAKOV 1986, pp.208 f), usually the basic assumption of local thermo-chemical equilibrium in the multicomponent system from the combustion chamber to the narrowest part of the nozzle is nothing more than a relatively comfortable and simplified but improved working hypothesis. In reality such a premise is questionable, considering the great variety of phenomena in non-equilibrium thermodynamic state. The calculations thus must be checked and verified in each individual case.

In order to clarify these circumstances satisfactorily, it is sufficient to mention three especially characteristic traits of such 'dissipative structures':

(1) Flows in rocket engines are characterized by gas dynamic parameters with, in part, extreme local gradients; the pressure is particularly affected, and often varies by several orders of magnitude.

(2) Due to the high combustion temperatures normally occuring within rocket engines, chemical reactions in non-equilibrium state often have substantial influence on the machine's performance.

(3) These chemical reactions sometimes lead to unexpected non-linear changes of the dominant influence parameters.

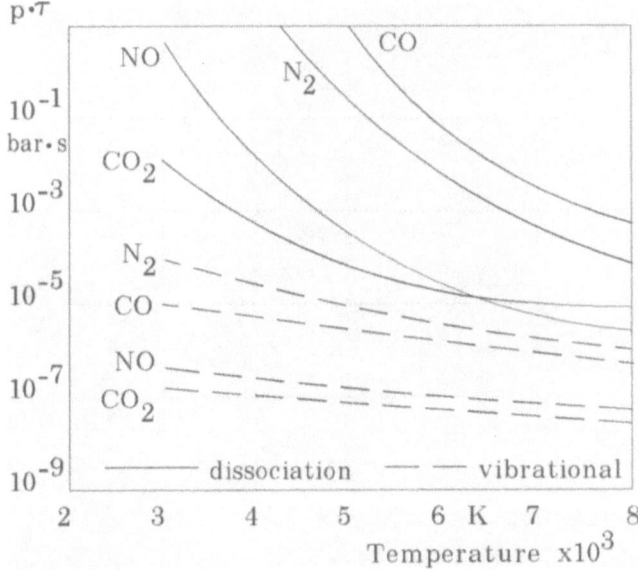

Figure 4.1: Characteristic Relaxation Times

The reason for this influence lies in the fact that "their relaxation times exceed, as a rule, those of the vibrational degrees of freedom of molecules by one or two orders of magnitude." (PIRUMOV & ROSLYAKOV 1986, p.209). The following diagram is taken from the same source.

In order to agree with the strong dependence on pressure, the product of pressure p and relaxation time τ is plotted above the temperature for some known diatomic gases. It is not difficult to see that the times lie, according to pressure, in orders of magnitude which are comparable with the time spent in the combustion chamber and Laval nozzle by the gases (see SUTTON 1986, p.201). Fortunately these complex relaxation processes can usually be adequately estimated for technical applications of a one-dimensional model (see PIRUMOV & ROSLYAKOV 1986, pp.213 f).[24]

5. Summary I

This study has two declared aims: it presents the theoretical basis for a provably ideal comparative process for relaxing flows (ICP) and justifies its application to jet and, in particular, rocket engines. This will be treated in two parts.

Part I offers a status quo report on current calculation methods, and compiles and explains briefly the most important data on selected prominent rocket engines. Starting from the phenomenology of the dynamical and physico-chemical conversion processes in the fuel-oxidizer fluid mixture and in the burned gases, the ideal thermodynamic comparative process is then derived – as a defined sequential change of states in the system. In order to render this comparative process readily understandable, it is first applied to an appropriate model gas using algebraic equations for all relevant parameters. This model gas undergoes energy conversion processes without forfeiting the simplicity of presentation typical of classical gas dynamics. Above all, examination of this model offers proof that it is generally impermissible to use, as is done in practice, the familiar isentropic equation for flow changes of state continuously propagated in flow tube theory.

Elementary calculations immediately indicate essential attributes which are also typical for relaxing, multicomponent, one-phase systems, such as the significant 'pressure drop phenomenon' or the establishment of the steady mass flow rate as an 'eigenvalue' of the comparative process. Their relevance to the RE theory is stressed. It is noted that the 'pressure drop' is closely related to the thermodynamic state of the fuel and oxidizer components before ignition.

Finally, reservations which must be observed when using the ICP in both the design and practical assessment of a rocket engine are individually presented and analyzed. In order to be able to use the ICP's mathematical relationships between the primary design data as a first approximation of the actual physical relations, one is forced to take into account the numerous dissipative effects in relaxing flows described in the last section of Part I.

"The mathematicians and the physics men have their mythology;
they work alongside the truth never touching it;
their equations are false but the things work."
                                    -Robinson Jeffers-

Part Two (II): Thermofluiddynamics of Rocket Propulsion

1. Problems with the NASA-Methods

1.1 Intentions of the NASA-Lewis Code

The title of the NASA-Lewis Code SP 273:

> "Computer Program for Calculation of Complex Chemical
> Equilibrium Compositions, Rocket Performance, Incident
> and Reflected Shocks,and Chapman-Jouguet Detonations"

contains, in itself, a complete list of contents. The largest and most
important part of the extensive work was based on the **documentation** of
the computer programs available at NASA in 1967, completed after twenty
years of research on the calculation of chemical equilibria primarily
in multicomponent gas mixtures.

In order to demonstrate the great range of use for the computer programs, the Code offers a number of examples in which chemical equilibriums play a significant role. To elucidate their physical background,
the Code offers, in addition to some special computational informations, a compilation of numerous briefly-described thermodynamic and
gas dynamic basic problems as well as important physical relationships.
More cursory information presented in the Code, such as the performance
data of a rocket engine, is not related to the equilibrium data calculated from the computer program. The degree to which, for example, certain deviations from the accepted physical conditions result from the

calculation procedures for the equilibrium values governing rocket performance, is not discussed. Assumption of a negligible gas velocity in
the combustion chamber is the standard case in the Lewis Code: the
equilibrium equations based on this assumption will be termed as the
'Adiabatic Flame Combustion' (AFC) method. This procedure, still used
world-wide, was modified as early as 1965 by Rocketdyne Division, since
the cross section area of the normally cylindrical combustion chamber
is generally by no means much larger than the nozzle throat cross section area (see YEDLIN 1965, and FARMER et al. 1966). In literature such
a process is termed as 'Finite Area Combustion' (FAC).

The FAC method first offered the possibility of estimating the pressure
drop inside the combustion chamber. Extensive performance calculations
showed, however, that this conversion effect did not provide a satisfactory explanation for the so-called 'combustion efficiency problem'
revealed in the AFC method (see PROZAN 1969, p.1). Even in the J-2 engine, with its unusually low area ratio A_C/A_T and a (reversible) pressure drop of nearly 20% inside the combustion chamber, the calculated
values of the specific impulse of the burned gases in the nozzle outlet
section still lie, with a given feed pressure and mass flow rate,
clearly above data established in experiments (see Table 1 in Part I).
If one would correct this physically obvious result by taking the mentioned 'pressure drop' into consideration and reducing the theoretical
thrust to the experimental with a corresponding reduction of the mass
flow rate, one may discover that the mass flow rate calculated with the
FAC method is clearly lower than the corresponding experimental value,
and the theoretical specific impulse remains, as before, too high (see
PROZAN 1982, p.4).

This situation is not surprising, insofar as one wishes to develop an
ideal comparative process for the real thermochemical reactions in a
rocket engine. Such an intention is expressed in the NASA-Lewis Code
section titled 'Assumptions' containing the conditions required for
calculating the various performance parameters of a rocket engine: "one
dimensional form of the continuity, energy, and momentum equations, ...
complete combustion; adiabatic combustion; isentropic expansion; homogeneous mixing; ideal gas law, ... For equilibrium performance, compo-

sition is assumed to attain equilibrium instantaneously during expansion..."(GORDON & McBRIDE 1976, pp.33-34). In fact, however, this theoretically important possibility of having an ideal comparative process available apparently has always been virtually ignored. **Of prime importance has always been the application of the Lewis Code for the simulation of real processes in rocket engines and its use as a planning instrumentarium and tool for calculating the properties of state not directly obtainable through experiments!**

This mixture of 'wish' (= computer program for evaluating a 'theory' whose premises are no longer known, or cannot be assessed, or are purposely ignored) and 'reality' (= 'experiment', which is doomed from the beginning to unfavorable design because of the missing comparative process) is amazingly effective. There are statements from renowned space scientists to the effect that deviations of more than 0.5% between experiment and theory (= NASA-Lewis Code) are inconceivable!

It is naturally quite logical that such deviations (with relatively low tolerances, as seen in the example mentioned above) between 'theory' and 'experiment' demand a plausible explanation, if for no other reason than to assure users of the established methods' reliability. This problem became particularly acute with the development of a reusable main rocket engine system for the space shuttle fleet. The requirement of **reusability** automatically made the cooled material zones – exposed to gas temperatures far above 3000 K when a LH-LOX mixture is burned – a major neuralgic point in the configuration. Without giving up the core of the NASA-Lewis Code – its famous equilibrium computer program – a fundamental review of the "Rocket Performance" theory was commissioned. Establishment of a realistic estimation of the gas temperatures in the nozzle throat not obtainable experimentally was one of the review's prime goals.

NASA first assigned this task to Lockheed (1969) and, considerably later (1982), to Continuum Inc. of Huntsville, Alabama. In both cases the problem was treated by R.J. Prozan. His results, documented in two contracts not yet officially available, were unanticipated. As one can see in Table 1 in Part I, his calculated thrust value coincided so well

with the test value that he arrived at the following optimistic conclusion: "The excellent comparison with the J-2 experimental evidence
indicates that for this case at least, finite rate, droplet vaporiza-
tion, mixing and boundary layer are secondary effects and the principle
source of error has been the misstatement of the equilibrium condition"
(PROZAN 1969, p.28). The visionary goal was apparently attained: the
real effects were revealed as only insignificantly different from the
Prozan's modifications of the Lewis Code. Thus a complete simulation of
the complex physico-chemical processes in a rocket engine was rendered
possible with a relatively simple theory. In addition, the new theory
predicted considerably lower and more manageable gas temperatures in
the nozzle throat cross section!

It is obvious that these 'perspectives' are extraordinarily attractive
in an era in which the shuttle engines are being subjected to ever
heavier loads and NASA is planning rocket engines for subsequent gener-
ations. Thus the Marshall Space Flight Center (MSFC) invited experts to
a workshop meeting in Huntsville, Alabama, at the end of February 1985.
Along with the staff experts of MSFC, R.J. Prozan, S. Gordon and a number of US experts from various universities, the author took part in
the conference as the only European, at the special request [25] of the
MSFC.

In a final communiqué unanimously passed by all the experts, Prozan's
results were rejected as unfounded. An extension of the Lewis Code was
deemed urgently necessary. It was recommended that the Munich Method be
investigated and, if found suitable, be used as the theoretical basis
for an "Extended Lewis Code".

In order to understand the quintessence of this decision, the fundamentals of the AFC and FAC methods as well as the **Prozan Procedure (PP)**
will be briefly reviewed below, together with their basic defects and
problem-specific approximations. Part II, however, is primarily devoted
to a presentation of the Munich Method and its physical verification.

1.2 AFC Method

Basis of all the NASA methods discussed below is the computer program for calculating chemical equilibrium in a predominantly gaseous, multi-component mixture, as documented in the NASA-Lewis Code. Since a discussion of the basic definitions of the chemical equilibrium play a decisive role in both the Prozan and Munich methods, a general introduction of the term is called for. The second edition of F.A. Williams' book (1985) is quite helpful in providing alternative viewpoints.

*   Chemical Equilibria

The traditional thermodynamic description of each equilibrium is based on the pertinent Gibbs function of the system: in this case, of the mixture of K components. In continuous systems, one normally assumes that enthalpy H as a Gibbs function, is known (see, for example, CALLEN 1966, p.99) to be dependent on the entropy S and the pressure p of the mixture as well as on the mole number n_k of all its species, known together as the **canonical properties of state**: $\hat{H} = H(S,p,n_k)$ for k=1(1)K. If these variables completely describe the behavior of the states of the system, then by using the total differential

$$(1.1) \qquad dH = TdS + Vdp + \sum_k \mu_k^m \, dn_k$$

one arrives at the temperature T, the volume V of the system as well as the (molar) chemical potential μ_k^m of the mixture's each component – known as the **conjugated** properties of state.

If the mixture undergoes a process realized by an isobaric change of state and at constant enthalpy, the mole number n_k^{equ} in equilibrium can be found through the following extremum condition

$$(1.2) \qquad \left[\frac{\partial S}{\partial n_k}\right]_{H,p,n_\ell} = 0 \quad ; \quad \ell,k = 1(1)K \quad ; \quad \ell \neq k \quad ;$$

derived from equation (1.1). As can be proven, the entropy S of the system is maximized.

For other process realizations — for example, in the standard case of an isobaric-isothermal combustion — another Gibbs function for describing the system is necessary: the **free enthalpy** G. One derives this property of state G from a **Legendre transformation** (see CALLEN 1966, pp.90 f), which converts equation (1.1) into the form

$$(1.3) \qquad dG = - \, SdT + Vdp + \sum_k \mu_k^m \, dn_k$$

and identifies G as dependent on the variables T, p and n_k (see SMITH & MISSEN 1982, p.42). In this form, the Gibbs function $\hat{G}(T,p,n_k)$ is defined by

$$(1.4) \qquad G := H - TS \qquad .$$

The chemical equilibrium — that is, the values n_k^{equ} at given unchangeable values for T and p — is now established by the extremum condition (equivalent to equation (1.2))

$$(1.5) \qquad \left[\frac{\partial G}{\partial n_k}\right]_{T,p,n_\ell} = 0 \quad ; \quad \ell, k = 1(1)K \quad ; \quad \ell \neq k \quad ;$$

it is well-known that the free enthalpy is minimized.

The close relationship between equilibrium conditions and realization of the process will be subsequently discussed in greater detail: for the moment emphasis is given to the definitive calculation of n_k^{equ} on the basis of equation (1.5). Two different conceptual possibilities are available: the stoichiometric or non-stoichiometric formulation of the equilibrium conditions.

The older **stoichiometric formulation** (see PENNER 1957, pp.146 & 149 f for H_2-O_2-combustion) transforms equation (1.5) into an expression which allows the establishment of the unknown mole numbers n_k^{equ} iteratively from the product sum

$$(1.6) \qquad \prod_k n_k^{\nu_{kr}} = \hat{K}_r(T,p \,|\, n_k^{ref}) \quad ; \quad k = 1(1)K$$

$$r = 1(1)\tilde{R}$$

when the given conditions for the reference state mole number n_k^{ref} are observed. Equation (1.6) is related to the K components of the mixture, whose chemical mechanism is described by $\tilde{R}$ linear independent chemical reactions; the integer condition $\tilde{R} < K$ is valid.

The known **equilibrium constants** of the r^{-th}-reactions are usually given with the expression $\hat{K}_r(T,p|n_k^{ref})$: it includes, parametrically, the mathematical dependence of pertinent n_k^{equ}'s on both the variables p and T . For ideal gases it is easily shown that the function $\hat{K}_r(T,p)$ may be factorized. There is particular interest in the equilibrium constant K_p:"the practical importance of K_p is the result of the fact that it is independent of total pressure and can therefore be listed as a unique function of temperature." (PENNER 1957, p.124).

The stoichiometric coefficients ν_{kr} are given numbers whose individual values are dependent on the reaction mechanism, as expressed by the **reaction equation**[26]

$$(1.7.1) \qquad \sum_k \mathfrak{C}_k \, \nu_{kr} = 0 \quad ; \quad k = 1(1)K \quad ; \quad r = 1(1)\tilde{R}$$

and which are combined in the stoichiometric matrix $\underline{\nu}$. Its rank

$$(1.7.2) \qquad \tilde{R} := \text{Rank} \, (\underline{\nu})$$

corresponds with the number of linear independent reaction equations. For calculating chemical equilibriums, the relationship

$$(1.7.3) \qquad E = K - \tilde{R}$$

between the total number K of all components (index k) active in the $\tilde{R}$ reactions (index r) and the total number E of the elements constituting these components (index e) is important. E is customarily equal to the rank of a matrix $\underline{A}$

$$(1.8) \qquad E = \text{Rank} \, (\underline{A}),$$

which is termed as **formula matrix**, and which, together with equation (1.7.2), leads to

$$(1.9) \qquad \underline{A} \cdot \underline{\nu} = \underline{0} \quad ;$$

$\underline{0}$ is the zero matrix.

The formula matrix contains as elements a_{ek} the same integers which state how many particles of the eth-element are present in all parts of a molecule or atom of the kth-component (for example, $a_{26} = 2$ for the element H (e = 2) in the component H_2O (k = 6) of a gaseous reactive system $\{H_2, H, O_2, O, HO, H_2O \mid O, H\}$ out of K = 6 components and E = 2 elements).

With the help of $\underline{A}$, the particle conservation (on the baryon level) can be expressed in the especially compact form:

(1.10)
$$\underline{A} \cdot n = b^{\emptyset} \quad .$$

This vector-matrix-form combines the component-quantity vector n with the element-quantity vector b for a reference state (index $\emptyset$).

If one takes one of the element-quantity equations out of Eq. (1.10)

(1.10.1)
$$\sum_{k=1}^{K} a_{ek} n_k = b_e^{\emptyset} \quad ; \quad e = 1(1)E \quad ,$$

n_k signifies the generally variable mole number of the kth-component and $b_e^{\emptyset}$ is the known mole number for the reference state of the eth-element's atoms in bounded and unbounded states (for example, unbounded as second component, or bounded in H_2O).

Equation (1.10.1) now appears as an additional constraint in the **non-stoichiometric formulation** of the equilibrium condition

(1.11)
$$minG(n) = min\sum_{k}^{K} \mu_k^m n_k$$

for the free enthalpy G of the system with given values of T and p. The reaction mechanism is bounded by equation (1.9).

The equilibrium values n_k^{equ} needed to fulfill equations (1.11), and (1.10.1) must be established. The solution to the constrained minimization function is facilitated by the familiar method of Lagrange multipliers with λ_e, $e = 1(1)E$, which are combined in the (transposed) vector $\lambda = (\lambda_1, \ldots, \lambda_E)^T$. They are defined by the **Lagrange Function**

(1.12)
$$\mathcal{L}(n, \lambda) := \sum_{k} \mu_k^m n_k + \sum_{e} \lambda_e [b_e^{\emptyset} - \sum_{k} a_{ek} n_k] \quad ,$$

$$k=1(1)K; \quad e=1(1)E;$$

which leads, with the instructions

(1.13)

$$\left[\frac{\partial\mathcal{L}}{\partial n_k}\right]_{n_\ell, \lambda} = \mu_k^m - \sum_{e} a_{ek} \lambda_e = 0$$

$$\left[\frac{\partial\mathcal{L}}{\partial \lambda_e}\right]_{\lambda_\ell, n} = b_e^{\emptyset} - \sum_{k} a_{ek} n_k^{equ} = 0$$

,

to a set of $(K + E)$ equations for the $(K + E)$ unknowns λ and n^{equ} (see SMITH & MISSEN 1982, pp.105 f). This elegant method is particularly transparent for mixtures of ideal gases, since the (molar) chemical potential μ_k^m of the k^{-th}-component in a 'closed' system refers exclusively to the mole number n_k of the k^{-th}-component:

$$(1.14) \qquad \boxed{\frac{\mu_k^m}{RT} = \frac{1}{RT} \hat{\mu}_k^{m\dagger} (T) + \ln\frac{p}{p^\varnothing} + \ln\frac{n_k}{n}} \qquad ; \quad k = 1(1)K.$$

Compare equation (2.e) of Appendix 2.

For process realization conditions T = constant and p = constant, the total mole number $n = \sum n_k$ in a given reaction volume V is established by the thermal equation of state for an ideal gas

$$(1.15)^{27} \qquad\qquad pV = n\ R\ T$$

so that the values μ_k^m appearing in equation (1.13) are variable only over the term $\ln(n_k)$. If the molar density n/V is variable, however, (as in 'open' systems), then μ_k^m depends on all mole numbers and a strong non-linear equation system arises. The second term of equation (1.14) depends, together with the known function $\mu_k^{m\dagger}$, only on the given values of T and p which thus play the role of process parameters.

The NASA-Lewis Code contains a solution algorithm of the second order, used for calculating the equilibrium mole numbers n_k^{equ} as a function of T and p. The basis for this algorithm is the non-stoichiometric formulation of the $G(n)$-minimization. The NASA algorithm consists of an efficient method for solving the often large and at times strong non-linear equation sets (1.13) and (1.14). It is the product of more than 15 years of development, dating back to 1947, when S.R. Brinkley Jr.'s first publication appeared (see WILLIAMS 1985, p.537). Brinkley's variation method was subsequently relegated to the background with the 1958 appearance of W.B. White's algorithm of the RAND Corporation (SMITH & MISSEN 1982,[28] p.127). The two procedures differ considerably from one another: the Brinkley algorithm uses the logarithm of the mole number n_k as a variable and satisfies the equilibrium conditions (1.13) with every iteration, whereas the RAND algorithm works directly with n_k and satisfies the additional constraint (1.10.1) with every iteration.

The development of the NASA algorithm was substantially formed by the notable contributions of S. Gordon, who worked on the first modifications in 1951 and subsequent alterations in 1959 and 1962. In the currently accepted version dating from 1976, the NASA iteration scheme also uses $\ln(n_k)$ as a variable; as a calculation procedure especially developed for non-linear equation sets, however, it avoids fulfilling the physical equilibrium conditions at each stage of iteration. Using the well-known Newton-Raphson-Procedure, all equilibrium mole numbers n_k^{equ} of the components and the Lagrange multipliers λ_e of the elements are obtained for the standard case (p, and T given). To do this, one iteratively executes the solution of the equilibrium relations (1.13) prepared for the numerical calculations and partly linearized (labeled as Newton-Raphson equations (18) to (23) in the Lewis Code).

* The AFC Method

The procedure for calculating the 'Adiabatic Flame Combustion' is still the most widespread and important method used for the NASA algorithm. The following sketch illustrates the technical conception upon which it is based.

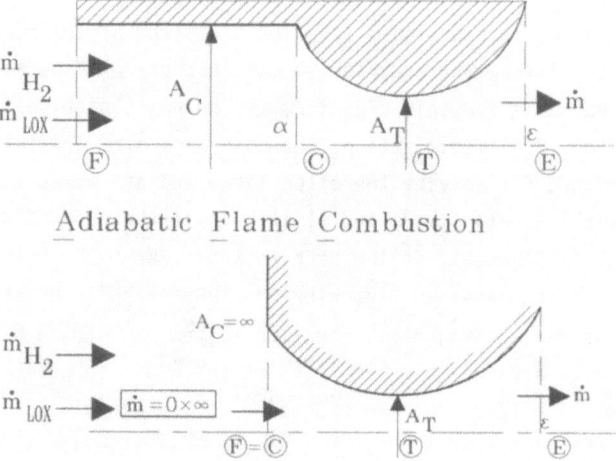

Figure 1.1: Basis Configuration for the AFC- and FAC Methods

The system (= burned gas) passes through the sequence of states $\underline{F}$, $\underline{I}$
and $\underline{E}$ along the axis of a Laval nozzle, which is attached to a combustion chamber with a cross section area A_C arbitrarily larger than the
nozzle throat cross section area. With this configuration $\alpha \rightarrow \infty$ holds,
states $\underline{F}$ and $\underline{C}$ coincide. This circumstance means that, strictly speaking, the steady mass flow rate $\dot{m}_T$ (determined by the nozzle throat
cross section area) in state $\underline{C}$ can be postulated but not verified.
Whether or not the real steady mass flow rate $\dot{m}_L$ (localized immediately
next to the entrance of the Laval nozzle) matches the determined value
$\dot{m}_T$ remains basically unknown. As a result of $A_C \rightarrow \infty$, the finite mass
flow rate in $\underline{C}$ is established through an undetermined expression $\dot{m} =$
$\rho(Av) \approx \rho(\infty \cdot 0)$, which probably assumes the value $\dot{m}_L < \dot{m}_T$ in that location. This is a fundamental ambiguity in the theory!

In the sketch, the AFC method differs from the FAC method only in one
single line. The horizontal line leading from $\underline{C}$ to $\underline{F}$ signifies a constant but finite (usually cylindrical) cross section area A_C for the
combustion chamber. In this significantly more realistic model, the two
states $\underline{C}$ and $\underline{F}$ no longer collapse.

A physically decisive difference is evident: in $\underline{F}$ one has the low temperature state of a fluid mixture composed of two liquid components, in
$\underline{C}$ the ignited high temperature gas mixture consists of six components.
In state $\underline{\infty}$ immediately behind the injector, the oxidizer and fuel flows
still haven't evaporated and mixed. In $\underline{F}$ there is, compared to the ρ-
value in $\underline{C}$, a mass density higher by the factor 100. Thus $v_F^2 \ll v_C^2$ is
valid at a constant combustion chamber cross section area, and a pres-
sure change occurs: $p_C < p_F$. A constant mass flow rate is assured, contrary to the limiting case $\alpha \rightarrow \infty$, since there are no conditions existing which correspond to an indefinite expression in the continuity
equation. Unfortunately there are serious problems (see Appendix 2) in
determining the necessary thermal and caloric data for the fluid mixture in $\underline{F}$. The specific entropy s_F, in particular, cannot be immediately calculated for this condition in the case of a LH-LOX mixture.

Both the AFC and FAC methods deal with a complete combustion chamber-
Laval nozzle configuration. Their names, however, reflect a far more
complex part of the procedure: determination of the **adiabatic flame**

temperature T_{ad} . It is calculated with the NASA algorithm explained above and serves as the initial value for calculating T_C. Prerequisites for using this algorithm will be briefly described because they affect the core of both Prozan's procedure and the Munich Method.

According to the First Law of thermodynamics, the enthalpy of a system remains unchanged in an adiabatic flow process. Insofar as isobaric combustion processes also simultaneously occur, pressure p and specific enthalpy h of the system are the properties of state to be kept constant for determination of the burned gas' mole number n_k^{equ} in equilibrium. These properties of state are (along with n_k), according to equation (1.1), the canonical variables for the specific entropy s

$$(1.16) \quad ds = T^{-1}dh - (T\rho)^{-1}dp - T^{-1}\sum_k \mu_k \, d\omega_k \quad ; \quad k = 1(1)K \ ,$$

and determine their extremum behaviour as equilibrium condition (see equation (1.2)). With specific values it is preferable to use the **mass fractions** ω_k for the species instead of the mole numbers n_k. Both values are closely interrelated through the algebraic expression

$$(1.17) \quad \boxed{ \omega_k := \frac{\rho_k}{\rho} = \frac{\psi_k}{\psi}\frac{n_k}{n} = \frac{\psi_k}{\psi} X_k \quad ; \quad k = 1(1)K }$$

In this expression, ρ_k is the partial density, ψ_k the molar mass and X_k the **mol fraction** of the kth-component. For the equilibrium formulation, the requirements for the element conservation are

$$(1.10.2) \quad \sum_{k=1}^{K} a_{ek} \, \omega_k/\psi_k = (\psi n)^{-1} b_e^{\varnothing} \quad ; \quad e = 1(1)E \ ,$$

(compare equations (1.10) and (1.10.1)).

The definition of the Lagrange Function $\mathcal{L}_s(\boldsymbol{\omega},\boldsymbol{\lambda})$ should comprise the fact that — contrary to equation (1.2) — although pressure p is given, the corresponding temperature T is not: instead, the specific enthalpy h of the system is a property which, in ideal gas mixtures, is dependent only on T (and the ω_k). In this case, the equation

$$(1.18) \quad h - h^{\varnothing} = 0$$

is introduced as an additional constraint and, analog to equation

(1.10.2), is provided with a supplemental Lagrange multiplier λ^E. The Lagrange function for the specific entropy s to be optimized is then

$$(1.19) \quad \mathcal{L}_s(\omega, \lambda, T) := -s + \sum_e \lambda_e \left[\frac{b_e^\varnothing}{\psi n} - \sum_k a_{ek} \frac{\omega_k}{\psi_k} \right] + \lambda^E (h - h^\varnothing)$$

$$e = 1(1)E ; \qquad k = 1(1)K .$$

As with equation (1.12), one meets the necessary criteria for determining the required properties of state in equilibrium with the differentiation of $\mathcal{L}_s(\omega, \lambda, T)$ with respect to the variable components of the vectors ω and λ as well as to T. When this is carried out,

$$(1.20) \qquad \frac{\partial}{\partial T} \mathcal{L}_s(\ldots, T) = -\frac{\partial s}{\partial T} + \lambda^E \frac{\partial h}{\partial T} ,$$

the result is apparently the **trivial** case

$$(1.21) \qquad \boxed{\lambda_{LC}^E = T^{-1}} \qquad ,$$

which, due to equation (1.4), allows the equilibrium treated here to be completely attributed to the equilibrium existing at a minimum value of specific free enthalpy g. In this case the equilibrium conditions (1.13) and the constraint (1.18) are simultaneously computed with use of the NASA algorithm. With the specific enthalpy of the system

$$(1.22) \qquad \boxed{h = \sum_k h_k(T) \, \omega_k} \qquad ; \quad k = 1(1)K$$

resulting from the thermodynamic formalism (see CALLEN 1966, pp.192 f) for ideal gas mixtures, one can iteratively calculate the properties of state T and ω_k^{equ} from the equations (1.13) to (1.15), (1.17) and (1.18). The concluding conditions for the mass fractions or mole fractions from equation (1.17)

$$(1.23.1) \qquad \boxed{\sum_{k=1}^{K} \omega_k = 1} \qquad (1.23.2) \qquad \boxed{\psi = \sum_{k=1}^{K} \psi_k \, x_k}$$

should naturally be observed.

If one needs the adiabatic combustion temperature T_{ad} as well as the corresponding mass fractions $\omega_{k,ad}^{equ}$ for the limit $\alpha \to \infty$, the following information is necessary ($\underline{\bullet}$ is chosen as the reference state $\emptyset$)

(1) the pressure p_F ($\approx p_\infty$ by assumption)

(2) the specific enthalpy h_F

(3) the molar masses as well as the (only temperature-dependent) specific enthalpy h_k and functions $\mu_k^{m\dagger}$ (as part of the molar chemical potential) for each component $k = 1(1)K$

(4) the mole number $b_{e\infty}$ of the elements.

While the information for (3) usually is taken from the common standard tables (for example, the JANAF Tables) or, in the case of simple substances, is calculated with adequate accuracy using the various resources of statistical thermodynamics, some remarks are needed for points (2) and (4). These two positions are as important for the practical application of the theoretical discussions documented in this study as they are for the use of the NASA-Lewis Code. In order to emphasize practical applicability, the remarks pertain only to an unburnt fuel-oxidizer mixture of liquid (or fluid) hydrogen (LH) and liquid oxygen (LOX) disposal at reference state.

Normally in a steady operation of a rocket engine, **steady mass flow rates** $\dot{m}_{LH}$ and $\dot{m}_{LOX}$ are fed into the combustion chamber. Since both fluids are diatomic, the required element ratio $b_{O\infty}/b_{H\infty}$ is expressed as

(1.24)
$$\frac{x_{LOX}}{x_{LH}} = \frac{\psi_H}{\psi_O} \frac{\dot{m}_{LOX}}{\dot{m}_{LH}}$$

by these rates and molar masses.

The closure condition

(1.23.3)
$$\sum_{k=1}^{K} x_k = 1 \quad ,$$

for the mole fraction x_k instantly yields the required information

$$(1.25) \quad b_{H\infty} = \left[1 + \frac{\psi_H}{\psi_O} \frac{\dot{m}_{LOX}}{\dot{m}_{LH}} \right]^{-1} \dot{n}_\infty$$

$$b_{O\infty} = \dot{n}_\infty - b_{H\infty}$$

about the mole number of the elements $b_{H\infty}$ and $b_{O\infty}$ (per time unit) in the reference state $\underline{\infty}$, insofar as their total mole number (per time unit) $\dot{n}_\infty$ is given. If the total steady mass flow rate $\dot{m} = \dot{m}_{LOX} + \dot{m}_{LH}$ is also known, one can, using the definition

$$(1.26) \qquad \boxed{\dot{n}_\infty := \dot{m}/\psi} \qquad ,$$

reduce $\dot{m}$ to a total mole number rate $\dot{n}_\infty$ of the elements. The inlet mole number rates $b_{e\infty}$, $e = 1(1)E$, thus are assumed as given parameters.

A precise calculation of the specific enthalpy $h_F \approx h_\infty$ of the fuel-oxidizer mixture is considerably more complicated. Generally both fluids are not at the same temperature level and thus do not meet the requirements for equation (1.22). In addition, one must assume that the LH-LOX mixture is a real mixture in a fluid state (see Appendix 2). Thus the usual basis for calculation

$$(1.27) \qquad h_F^m = h^m(LH) \, x_{LH} + h^m(LOX) \, x_{LOX}$$

by no means satisfies the requirements of equation (1.22), but rather the First Law of thermodynamics for steady bulk flows. Referring to equation (2.a) of Appendix 2, one evaluates it simply with the use of the given molar enthalpies $h^m(LH)$ and $h^m(LOX)$ of both fluids at their different liquid temperatures (see PRIGOGINE & DEFAY 1962, p.58).

Using equations (1.24) and (1.23.3), a relationship with practical significance results from equation (1.27)

$$(1.28) \quad \boxed{h_F^m = h^m(LOX) + [h^m(LH) - h^m(LOX)] \left[1 + \frac{\psi_H}{\psi_O} \frac{\dot{m}_{LOX}}{\dot{m}_{LH}} \right]^{-1}} \quad ,$$

which, at a given **mass flow ratio** $\dot{m}_{LOX}/\dot{m}_{LH}$, immediately yields the molar enthalpy of the fluid mixture.[29]

With fixed values in **C** known from the AFC method, all properties of state in **I** and **E** can be successively determined from the NASA-Lewis Code as follows:

□ properties of state in the nozzle throat section **I**

(1) For determination of the pressure, an approximate value is calcu- lated for both **I** and the nozzle outlet section **E** . Without explana- tion, the formula

$$(1.29) \qquad p_T/p_C = [(\kappa_S + 1)/2]^{-\kappa_S/(\kappa_S - 1)}$$

is recommended (GORDON & McBRIDE 1976, p.37), along with the note: "The program uses the value of κ_S from the combustion point." The equilibrium program given above then supplies the equilibrium mole fractions $\chi_{k,T}^{equ}$ of the gas mixture, together with the pertinent tem- perature T_T in the 'throat'.

(2) With the known data in **C**, the specific entropy s of the mixture is also established. Using equation (1.22), it can be expressed, like the specific enthalpy h, as the weighted sum

$$(1.30) \qquad s = \sum_k s_k\, \omega_k \quad ; \qquad k = 1(1)K$$

of the **specific partial entropies** s_k **of every component**. In con- trast to the h_k, the s_k are not only dependent on the temperature T, but also on the pressure and (in a mixture of ideal gases) the mole fraction χ_k of the **assigned** component k:

$$(1.31) \qquad s_k = \hat{s}_k^\dagger (T) - (R/\psi_k)\, \ln(\chi_k p/p^{\varnothing}) \quad .$$

The $s_k^\dagger$ are pure temperature functions available either in tables or calculated with the help of statistical thermodynamics.

Using definition (1.17), equations (1.30) and (1.31) yield the spe- cific entropy s of the mixture for the prescribed values of $\chi_{k,C}^{equ}$ and T_C at the known pressure p_C. With an **isentropic** change of state

$$(1.32) \qquad \hat{s}(T,p,\chi_k^{equ}) = \hat{s}_C(T_C,p_C,\chi_{k,C}^{equ}) \quad ,$$

equation (1.32) gives a second relationship, in order to gain an improved value for the pressure p_T out of the first approximate values from point (1). In order to obtain the final connection between p_T, T_T and $X_{k,T}^{equ}$, this iteration (using other empirical approximate formulas recommended for p_T) is governed in the AFC method by the familiar constraint

(1.33.1) $$M_T = 1$$

underlying the choked mass flow rate for any initial state conditions.

(3) This 'Mach number condition' is expressed in the NASA-Lewis Code by the **convergence test**

(1.33.2) $$\left| \frac{v^2 - a^2}{v^2} \right|_T \leq 0,4 \ 10^{-4}$$

in which the local speed of sound a is determined by the temperature in the nozzle throat section with $(\kappa_S R \ T)^{\frac{1}{2}}$. The flow velocity v in $\underline{T}$ then results from the continuity equation $\dot{m} = A_T \ \rho_T \ v_T$. For its evaluation, the mass density ρ_T is obtained with the thermal equation of state (1.15) and equation (3.10 - I) respectively, and $\dot{m}$ as well as A_T are assumed as given.

Only when equation (1.32) and condition (1.33.2) can be simultaneously fulfilled, can the calculation for $\underline{T}$ be completed.

□ Properties of state in the nozzle exit section $\underline{E}$

With the AFC method, the user may specify as a known property either the pressure ratio p_C/p_E or the area ratio A_E/A_T . In the latter case, the NASA-Lewis Code again uses a number of empirical formulas, constructed like

(1.34) $\ln(p_C/p_E) = \kappa_S + 1,4 \ \ln(A_E/A_T)$ für $A_E/A_T \geq 2$,

in order to obtain a first approximation for an iterative calculation of the properties of state. In addition, as in the case of the nozzle throat section, sufficient information is necessary: the energy equation (for T_E), continuity equation (for v_E), thermal equation of state (for ρ_E), the isentropic condition (1.32) and the previously-mentioned equilibrium program for the determination of the pressure p_E and the $X_{k,E}^{equ}$. The iteration is, however, rather dubious, as is shown for a specific example in Section 1.5.

In **Figure 1.2** the governing equations for the 'isentropic expansion' along the nozzle (previously explained under the subtitle 'comparative process')are once again presented for the sake of clarity. The given iterative requirements for p_E will be more fully discussed in Sec. 1.5.

	Simulation Process (T) (E)		Comparative Process (T) (E)	
Temperature	$T = \dfrac{2}{\kappa+1} T_C$	$T + (\frac{T}{p})^2 R \frac{\kappa-1}{2\kappa}(\frac{\dot{m}}{\varepsilon A_T})^2 = T_C$	$h + \frac{1}{2}\gamma R T = h^{tot}$	$h + \frac{1}{2}v^2 = h_C^{tot}$
Pressure	$p = p_T (T/T_T)^{\frac{\kappa}{\kappa-1}}$	$p = p_C (T/T_C)^{\frac{\kappa}{\kappa-1}}$	$s(T,p,\chi_k^{equ}) = s_C$	$P_{j+1} = P_j \left(\frac{\varepsilon_{j-0}}{\varepsilon}+j\frac{\Delta\varepsilon}{\varepsilon}\right)^{\alpha_j}$
Mass Flow Rate	$\dot{m} = \dfrac{A_T}{\rho_C}\sqrt{\kappa p_C}\left(\frac{2}{\kappa+1}\right)^\delta$	$\dot{m}_E = \dot{m}$	given	$\dot{m}_E = \dot{m}_T$
Mach Condition	$\left|\frac{v^2-a^2}{v^2}\right| < \beta_s$	not applicable	$\left|\frac{v^2-a^2}{v^2}\right| < \beta_{v_1}$	not applicable
Isentropic Condition	not applicable	not applicable	not applicable	$s(T,p,\chi_k^{equ}) - s_C < \beta_{v_2}$
Isentropic Exponent	Iterationvariable $\kappa_{l=0}= \gamma_C$	$\kappa = \kappa_T$	not applicable	not applicable
Velocity	$v = \dfrac{\dot{m}}{\rho A}$		Speed of Sound	$a = (\gamma RT)^{\frac{1}{2}}$
Mole Fraction	min $G(T,p)$		Density	$\rho = \dfrac{p}{RT}$

Figure 1.2: Isentropic Nozzle Expansion Flow for Comparative or Simulation Processes $\qquad \delta = \dfrac{\kappa+1}{2(\kappa-1)}$

The FAC method, incidently, is often employed in preparation for classical gas dynamics calculations. Then the NASA-Lewis Code, with values prescribed for the reaction enthalpy h_F, feed pressure p_F and mass flow ratio, offers at first all properties of state in **C**.

As an example, the calculation scheme used for the subsequent determination of the properties of state in **I** and **E** is shown in Figure 1.2 under the column 'simulation process'. It is obvious that this procedure no longer can claim to prove the intended isentropic changes of state. As shown in Chapter 3 of Part I, the simple isentropic relationships of type $p\rho^{-\kappa}$ = constant no longer are valid in processes with conversion events. They are computing rules for simulation of a dissipative flow of the burned gases within a Laval nozzle-shaped channel. Their use can be justified only heuristically. Nevertheless the calculation scheme contains -in contrast to the NASA-Lewis Code – a requirement for the determination of the steady mass flow rate based on Bray's gas dynamical criterion (compare the correction function Ω_m in equation (3.17) of Part I).

1.3 FAC Method and Prozan's Procedure

Figure 1.1 clearly reveals the intentions of the **Finite Area Combustion method**: engines have a combustion chamber with a finite cross section area A_C . In addition, the ratio of A_C to the nozzle throat cross section area A_T often lies under 2. An energy conversion generally occurs within the system due to this finite section. A considerable pressure drop results at both the combustion chamber end and the entrance to the subsequent nozzle even when dissipative effects are disregarded. Along with the corresponding increase of the flow velocity of the burned gases in $\underline{C}$ compared with $\underline{F}$, certain consequences arise which seem to make the use of the AFC method problematic from the beginning, at least for the 'nozzle calculation' under the label 'isentropic expansion'.

Past experience has shown that the reversible nature of the pressure drop should be emphasized. It must not get confused with dissipative losses in static and total pressure respectively.

The FAC method is simply an extension of the AFC method. Its primary aim is the calculation of the properties of state in the combustion chamber section $\underline{C}$ using the balance equations of the flow tube theory (Chapter 3 in Part I). Here they are combined in a form commonly used in the FAC method:

(1.35)
$$p + b\,\nu = p_F$$

(1.36)
$$h + \frac{b}{2}\,\nu^2 = h_F$$

The abbreviations (see PROZAN 1969, p.9)

(1.37) $\qquad b := (\dot{m}/A)^2 \quad ; \quad \nu := (RT/p\psi)$

refer, just like pressure and specific enthalpy, to the state $\underline{C}$; they are supposed to imply that the steady mass flow rate $\dot{m}$ is considered as a freely predictable value, and b represents an operational parameter with a known cross section area ($A = A_C = \alpha\,A_T$).

First the specific enthalpy h_F of the fluid mixture is determined (using equation (1.28)) for the given parameters p_F, $\dot{m}$ and $\dot{m}_{LOX}/\dot{m}_{LH}$ (in the case of a LH-LOX combustion); the input values for T_{ad} and $X_{k,ad}^{equ}$ are subsequently calculated with the AFC method.

One can thus compare iteratively the two unknown values of the thermal properties of state p and T in $\underline{C}$, considering that the specific enthalpy h of an ideal mixture is only dependent on T and X_k through equation (1.22). Care must be taken to use a consistent set of units.

A decisive precondition for this procedure is the inclusion of the equilibrium calculation (given above) for the $X_{k,C}^{equ}$, with actual T and p values in the iteration sequence. When p_C, T_C and the $X_{k,C}^{equ}$ are thus obtained, the mean molar mass ψ_C can be derived from $\psi_C = \Sigma \psi_k X_{k,C}^{equ}$, the mass density ρ_C of the gaseous mixture from the thermal equation of state and, finally, the flow velocity v_C from the condition for constant mass flow density $\rho_C v_C = \dot{m}/A_C = \sqrt{b}$ in the combustion chamber.

With the properties of state in $\underline{C}$ now known, the 'isentropic expansion' calculations presented above are used in the FAC method to go from $\underline{C}$ through $\underline{I}$ to $\underline{E}$ (appropriately enough, however, only with the variants presented under the column labeled 'comparative process' in Fig. 1.2). This point should be noted, since the **Prozan Procedure** to be discussed here is not focused directly on this part of the FAC method.

Prozan did not mention that the set of equations from (1.35) to (1.37), including the thermal equation of state, is mathematically overdetermined: the Earnshaw Paradox results (see BIRKHOFF 1950, pp.22 f) if these balance equations for a combustion chamber flow are explicitly supplemented by an appropriate reversibility condition similar to equation (1.32). Prozan's concern is with the question of whether and, if so, how the problem of the chemical equilibrium presented above is affected with regard to the usual gasdynamical equations of motion and energy, as given for a duct with a finite section and continuously running processes.

Prozan's question undoubtedly touches the core of the problem! His major premises thus should clearly be recapitulated:

(i)   The standard method for calculating the equilibrium composition of a chemically-reacting mixture is the Gibbs free energy minimization procedure, with p = constant and T = constant as given process conditions.

(ii) For technically relevant combustion calculations, the unknown combustion temperature is of interest. Thus the operational conditions for the process realization with negligible parts of the local kinetic energy now are – in contrast to (i) – p = constant and h = constant. If one assumes that the trivial case (1.21) applies to the corresponding Lagrange multiplier (see also PROZAN 1969, p.10 f), (ii) apparently is usually attributed to (i).

(iii) In the case of Finite Area Combustion, the physical facts are such that the constancy of neither the pressure p nor the specific enthalpy h is a realistic condition for process realization: only their total values remain constant in an isentropic flow.

Prozan's meritorious attempt at a solution is particularly notable because, although his diagnosis is correct, his therapy is wrong. Unfortunately the matter is so complex that it requires a number of explanations based principally on the thermodynamic equilibrium concepts discussed in the previous section.

Analog to the previously-treated two functions of state $\hat{G}(T, p, n_k)$ and $\hat{s}(h,p,n_k)$, a new canonical function $\hat{Y}(\mathbf{X}^{can}, \mathbf{n})$ is introduced which defines the vector $\mathbf{X}^{can}$ wholly with A variables. Such functions of state are dependent on their canonical variables, and their extremum values concerning the mole number n_k (the other canonical variables are to be fixed) describe chemical equilibrium.

If we choose the non-stoichiometric formulation of the equilibrium condition, in a generalization of equations (1.12) and (1.19) two typical versions of the calculation procedure for the unknown equilibrium values are encountered:

(1)
$$\mathcal{L}(\mathbf{n},\boldsymbol{\lambda}) := Y + \sum_e \lambda_e (b_e^{\emptyset} - \sum_k a_{ek} n_k)$$

$$\ell,e := 1(1)E \; ; \; \ell,k := 1(1)K$$

(1.38)
$$\left[\frac{\partial \mathcal{L}}{\partial n_k}\right]_{n_\ell, \boldsymbol{\lambda}} = \left[\frac{\partial}{\partial n_k} Y\right]_{\mathbf{X}^{can}} - \sum_e \lambda_e a_{ek} = 0$$

$$\left[\frac{\partial \mathcal{L}}{\partial \lambda_e}\right]_{\lambda_\ell, \mathbf{n}} = b_e^{\emptyset} - \sum_k a_{ek} n_k^{equ} = 0$$

$$X_\alpha^{can} = \text{constant} \; ; \quad \alpha = 1(1)A$$

(2)

$$\mathcal{L}(\mathbf{n},\lambda,T,p) := Y + \sum_e \lambda_e(b_e^\emptyset - \sum_k a_{ek}\, n_k) + \sum_\beta \lambda_\beta(X_\beta^{can} - X_\beta^\emptyset)$$

$$\beta := 1(1)B$$

$$\left[\frac{\partial\mathcal{L}}{\partial n_k}\right]_{n_\ell,\lambda,T,p} = \left[\frac{\partial}{\partial n_k}\, Y\right]_{X^{can}} - \sum_e \lambda_e\, a_{ek} = 0$$

$$\left[\frac{\partial\mathcal{L}}{\partial\lambda_e}\right]_{\lambda_\ell,n,T,p} = b_e^\emptyset - \sum_k a_{ek}\, n_k^{equ} = 0$$

(1.39)

$$\left[\frac{\partial\mathcal{L}}{\partial T}\right]_{n,\lambda,p} = \sum_\beta \left[\left(\frac{\partial Y}{\partial X_\beta^{can}}\right)_{n,X_{\neq\beta}^{can}} + \lambda_\beta\right]\left[\frac{\partial X_\beta^{can}}{\partial T}\right]_{n,p} = 0$$

$$\left[\frac{\partial\mathcal{L}}{\partial p}\right]_{n,\lambda,T} = \sum_\beta \left[\left(\frac{\partial Y}{\partial X_\beta^{can}}\right)_{n,X_{\neq\beta}^{can}} + \lambda_\beta\right]\left[\frac{\partial X_\beta^{can}}{\partial p}\right]_{n,T} = 0$$

$$\left[\frac{\partial\mathcal{L}}{\partial\lambda_\beta}\right]_{\lambda_\ell,n,T,p} = X_\beta^{can} - X_\beta^\emptyset = 0$$

$$X_\alpha^{can} = \text{constant}\;;\qquad \alpha \neq \beta\;;$$

Vector $\mathbf{X}^{can}$ of the canonical variables X_α^{can} ($\alpha = 1(1)A$) also includes those (index ß) which can be expressed as functions of the thermal properties of state; $\hat{s}(h,p,n_k)$ offers an example each for $X_\alpha^{can} = p$ and for $X_\beta^{can} = h$.

The two Lagrange functions (α) and (ß) are true for two different mathematical problems:

□ In the G-case, the mole numbers n_k^{equ} in chemical equilibrium are calculated from the Lagrange function $\mathcal{L}_Y^1$ (together with the uninteresting Lagrange multipliers $\boldsymbol{\lambda} = [\lambda_1,\dots,\lambda_E]$), with respect to the constraint relations $X_\alpha^{can} = \text{constant}$ ($\alpha = 1(1)A$); the X_α^{can} are the canonical variables of $Y(n_k)$; the equilibrium state is realized by the invariance of these canonical properties of state.

□ The same is true in the s-case as in the G-case, with one significant difference: here the definition of the Lagrange function $\mathcal{L}_Y^2$ with the same constraint relations as for $\mathcal{L}_Y^1$ is supposed to allow, in addition, the simultaneous calculation of temperature and pressure values belonging – along with the canonical variable X_α^{can} – to the n_k^{equ} for all the species in equilibrium.

The relative simplicity of the G-case is illustrated by the fact that T and p are in fact the canonical variables of the free enthalpy G(**n**).

Prozan's work consisted of finding the corresponding Lagrange functions for the governing balance equations (1.35) to (1.37). This could not be solved with conventional methods (see Chapters 2 and 3).
Even within the framework of conventional thermodynamics it is evident that a close relationship exists between $\mathcal{L}$, Y and $\mathbf{X}^{can}$ with regard to the thermodynamic equilibrium.

The stability of such an equilibrium condition is primarily guaranteed by the existence of a corresponding Ljapunov functional; sucha typical functional, however, can only be defined within an inherently consistent system of thermodynamic variables – just the canonical variables – (see STRAUB 1987). Prozan ignored this fact! He interpreted the Lagrange function $\mathcal{L}_Y^2(\mathbf{n},\lambda,T,p)$ for the mixture's (negative) specific entropy (s := –Y) as a function of both the thermal properties T and p , and the mole number vector **n**. According to scheme (2), this Lagrange function should be subject to all possible physically-required constraints X_β = constant.

Prozan then claimed that the basic problem of rocket engines (the constellation of a combustion chamber with a finite section and an adjoining Laval nozzle) is caused by the **feedback** of the chocking condition in the nozzle throat section on the combustion chamber flow. In addition to the governing balance equations (1.35) to (1.37), therefore, a supplementary condition must be taken into account: "the throat will act as a constraint on the combustion process, i.e.,

$$(1.40) \qquad\qquad \Phi^T := \ln(A^*/A_T) \equiv 0$$

where A^* is the chocking area resulting from an isentropic expansion away from the post-combustion condition." (PROZAN 1968, p.13).
This hypothesis was adopted by Prozan to define a Lagrange function $\mathcal{L}_\gamma^2(\mathbf{n},\lambda,T,p) := \mathcal{L}_s$ subsequently used to derive a set of equations, completely analogue to scheme (2). Upon solving this set, one may obtain all the properties of state at the exit of the combustion chamber:[30]

Prozan's corresponding Lagrange function is:

$$\mathcal{L}_s := -s + \sum_e \lambda_e (b_e^\emptyset - \sum_k a_{ek} n_k) + \lambda^E (h + \frac{b}{2} v^2 - h_F) +$$

$$+ \lambda^M (p + b v - p_F) + \lambda^T \Phi^T$$

$$e = 1(1)E \; ; \; k = 1(1)K \; .$$

(1.41)

$$\left[\frac{\partial \mathcal{L}_s}{\partial n_k}\right]_{n_\ell,\lambda,T,p} = -s_k - \sum_e \lambda_e a_{ek} + \lambda^E [h_k - bv^2 \psi_k(1 - \chi_k)] -$$
$$- \lambda^M bv\psi_k(1 - \chi_k) + \lambda^T(\partial\Phi^T/\partial n_k) = 0$$

$$\left[\frac{\partial \mathcal{L}_s}{\partial \lambda_e}\right]_{\lambda_\ell,\mathbf{n},T,p} = b_e^\emptyset - \sum_k a_{ek} n_k = 0$$

$$\left[\frac{\partial \mathcal{L}_s}{\partial T}\right]_{\mathbf{n},\lambda,p} = (\lambda^E T - 1)\frac{\partial h}{\partial T} + \lambda^E bv^2 + \lambda^M bv + \lambda^T(\partial\Phi^T/\partial T) = 0$$

$$\left[\frac{\partial \mathcal{L}_s}{\partial p}\right]_{\mathbf{n},\lambda,T} = \frac{R}{\psi} - \lambda^E bv^2 + \lambda^M(p - bv) + \lambda^T(\partial\Phi^T/\partial p) = 0$$

$$\left[\frac{\partial \mathcal{L}_s}{\partial \lambda^E}\right]_{\lambda_\ell,\mathbf{n},T,p} = h + \frac{b}{2}v^2 - h_F = 0$$

$$\left[\frac{\partial \mathcal{L}_s}{\partial \lambda^M}\right]_{\lambda_\ell,\mathbf{n},T,p} = p + bv - p_F = 0$$

$$\left[\frac{\partial \mathcal{L}_s}{\partial \lambda^T}\right]_{\lambda_\ell,\mathbf{n},T,p} = \ln(A^*/A_T) = 0$$

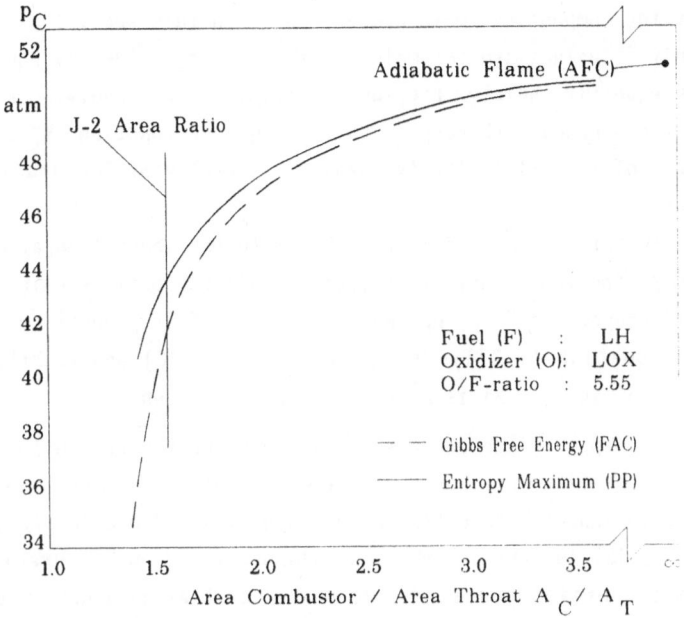

Figure 1.3: Combustor Exit Pressure p_C vs Area Ratio

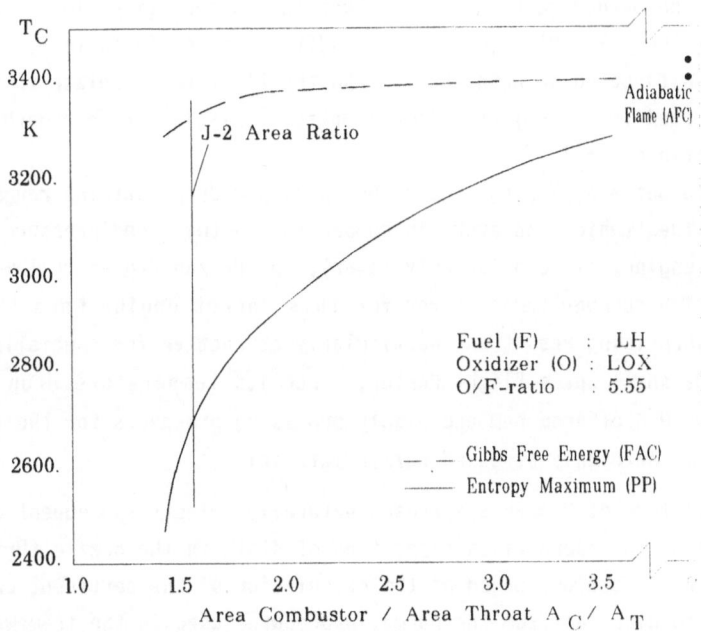

Figure 1.4: Combustor Exit Temperature vs Area Ratio

The equation set (1.41) allows a solution using mole number rates $\dot{n}_k$ instead of the components' mole numbers n_k. In this special case the relevant initial values are the mole number rates $b_{e\infty}$ given by equation (1.25). Consequently, solely Lagrange multipliers λ_e involving a time unit now are concerned with respect to the Lagrange function $\mathscr{L}_s^\beta$. This formal change of the multipliers's meaning is irrelevant for practical applications.

It is evident that system (1.41), together with the thermal equation of state (for ρ) and the 'constant' b (for v), allows complete calculations of all properties T, p, n_k and $n = \Sigma n_k$ in $\underline{C}$ if ρ equals v^{-1} according to equation (1.37). Using equations (1.30) and (1.31), the isentropic condition (1.32) is necessary for comparison.

In addition, an appropriate expression is known for the **Prozan function** Φ^T. General theoretical references about Φ^T do not exist: thus Prozan chose a relation known from classical gas dynamics and previously given as equation (3.13) in Part I. As noted there, it is not correct for gases with conversion processes. If one uses it as an approximation, one can easily check the effects of Prozan's hypothesis (1.40). In the two above <u>Figures 1.3 and 1.4</u>, calculated pressure- and temperature-values of the burned gases in the combustion chamber (i.e. in $\underline{C}$) are plotted for the J-2-engine (under the conditions given in Table 1, Part I). The results based on both the AFC and the FAC methods relate to the reference case ($\alpha = 1.58$) as well as for its variations with regard to the area ratio $\alpha = A_C/A_T$.

The differences are glaring: while the 'pressure drop' attains roughly matching values (which, however, in comparison to the feed pressure p_F in the J-2 engine, lie considerably lower), the Prozan Procedure lowers the combustion chamber temperatures for this rocket engine more than 20%. This surprising result was particularly attractive for specialists of NASA's Marshall Space Flight Center, since its temperature drop of more than 600 K offered new and highly promising prospects for the re-usability of the highly-stressed nozzle materials.

The special form of Prozan's approach naturally affects subsequent determination of the appropriate properties of state in the nozzle throat and nozzle exit section, based on the calculation of the pertinent data in $\underline{C}$. It also diverges from the normal procedures used in the framework of the AFC and FAC methods. This is not the place, however, for a detailed examination of Prozan's concept (see PROZAN 1982).

1.4 Evaluation of the NASA Methods

Criticism of the three calculation methods previously described in this study is increasingly justified as the standards applied to their results become more rigorous. The methods, in fact, move beyond the scope of scientific evaluation when the physical laws as well as the problem-oriented conditions for process realization are 'forgotten', and the entire routine degenerates into simple 'matching procedures'. Astonishingly enough, precisely this type of application of the NASA-Lewis Code is common practice today – especially in conjunction with experimental programs for rocket engines. This has previously been commented on in Sections 1.1 and 1.2 (see Figure 1.2). Examples of this practice will be presented below.

When, however, one takes the theory's original intentions seriously (to offer an **isentropic flow tube theory with chemical conversion processes**), there are compelling reasons for criticism. In order to provide a general survey, the most important defects of all these NASA procedures are briefly summarized in a table. Before doing so, however, some comments are given on the Prozan procedure, since it offers an especially good illustration of the difficulties encountered in formulating an effective evaluation system.

Prozan's solutions are derived from a set of equations which agree completely in form with the general non-stoichiometric equilibrium formulations of thermodynamics (equation (1.39)). This appearance, however, is deceptive: hypothesis (1.40) proposed by Prozan is by no means formulated using a variable provably being a **canonical variable** of the mixture's specific entropy which constitutes the Lagrange function $\mathcal{L}_s$; it has to arise in a Gibbs fundamental equation (GFE) of the same type as equation (1.1).[31] In such a GFE every additional independent variable should constitute an appropriate energy form, together with its conjugated property of state. **Such an energy form, however, is not known for the Prozan function Φ^T.** This fact accounts for the great difficulty in finding such an expression which can be adequately justified.

Only one conclusion can be drawn from this point of view: **the presented values derived from the Prozan procedure for the description of the system state C by no means represent a chemical equilibrium!**

The extent to which the Prozan procedure merits interest as a methodic calculation routine is not open to debate here. The procedure is, however, an especially clear illustration of the fact that, in physics, the internal basic relationships between a theory's independent and dependent variables and parameters as well as their physical significance are often more important than the theory's mathematical structure.[32] This is why the recommended NASA-Lewis Code procedures for calculating isentropic expansion in the Laval nozzle are also incorrect. Isentropic condition (1.32) cannot be precisely calculated, since the mole or mass fractions of the species occurring in the nozzle are not known; the normal practice of employing equilibrium values from the minimized free enthalpy at constant pressure and temperature values is theoretically inadmissible. This is not alone due to the nozzle expansion flow not being isothermal-isobaric. This fact is later examined in Section 3.5.

The following Table 1.1 lists various defects of the three discussed NASA methods with regard to characteristic features of the theory. Their evaluation is based on the three categories

  □ feature heeded and realized   (0)

  □ feature heeded, yet not realized   (∅)

  □ feature not heeded   ($)

Generally speaking, none of the three results satisfies expectations. Particularly serious is the virtually incomprehensible fact that all procedures treat the steady mass flow rate per area unit as a freely-given parameter and not as a resulting 'eigenvalue' (see the conclusion of Section 1.2). Although this required physical condition is obvious even in simple gas dynamical models (see Chapter 3 in Part I) and was discussed thoroughly as early as 1970 by Bray with regard to relaxing nozzle flows, it is not even mentioned in later editions of the NASA-Lewis Code. This is possibly explained by the fact that the performance calculations executed with the original NASA report instructions deal only with an infinite combustor cross section area.

Upon reviewing the defects of each of the three NASA methods, a fundamental dilemma is readily apparent: **their insufficient theoretical
transparency ultimately renders them obscure**. And, as will be shortly
demonstrated, they induce truly astounding irrationality when practically applied in a relevant computer program.

feature method	AFC	FAC	PP
p and h in fluid state $(\underline{F})$	O	O	O
chem. equilib.: chamber exit $(\underline{C})$	O	∅	∅
chem. equilib.: throat $(\underline{I})$	∅	∅	∅
chem. equilib.: nozzle exit $(\underline{E})$	∅	∅	∅
isentropic condition: max $s_F(\mathbf{n})$ $= s_E[.. \vert extr.Y(\mathbf{n})]$	∅	∅	∅
Mach number condition: $M_T = 1$	O	O	∅
nozzle exit condition: $p_E \geq p_U$	$	$	$
$\dot{m}$ as a 'eigenvalue'	$	$	$

<u>Table 1.1</u>: Basic defects of the NASA performance cal-
culations

To avoid misunderstandings it must be added that algorithms naturally
may be declared as obligatory. For example, one can regard the isentropic condition (1.32) simply as a separate iterative rule without taking
the entire physical connections into consideration. If this is done,
however, the resulting pressure and temperature values cannot be obtained using scheme (1.39) to establish the equilibrium composition
n_k^{equ} with problem-specified constraints. Thus they do not generally
fulfill the equilibrium coordinates. Yet such data is frequently considered adequate for practical applications, in spite of insufficient
comprehension!

The task of constructing a Lagrange function $\mathcal{L}_Y^2$ for scheme (1.39) is naturally anything but trivial. Yet it is the only way to find a thermodynamically consistent, inherently non-contradictory solution. All other solutions are approximations. They are only permissible and reasonable as long as one remains aware of their provisional character and knows that their dependence upon one of the numerous variables is always weighted with risks.

In conclusion it can be stated that the **condition for isentropic changes** ('isentropic' condition for the specific entropy of the mixture) at each point in the sequence of states

$$(1.42) \quad \boxed{\begin{aligned} s[T,p,n_k] &= s[T,p\,|\,extr.Y(\mathbf{n})] \\ &= s_F = s[T_F,p_F\,|\,minG(\mathbf{n})] \end{aligned}}$$

is not strictly fulfilled by any of the NASA procedures. Each procedure has its specific manner of approximating equation (1.42). As the results of a comparison between the PP and the AFC and FAC methods clearly show, the discrepancies can be considerable. The fundamental problem apparently lies **in finding the extreme value of the problem-specified canonical function of state Y with regard to n_k, and in simultaneously calculating the associated equilibrium pressure and temperature values,** using scheme (1.39).

1.5 Matching Procedures

These theoretical discrepancies in the NASA methods produce, in practical applications of the NASA-Lewis Code, unexpected ambiguities in the
results. This is illustrated below using data currently being discussed
for the development of the ARIANE 5-HM60/1 engine with H_2-O_2-combus-
tion.

The most important parameters[33] are presented in <u>Table 1.2</u>

feed pressure	p_F = 100 bar
nozzle throat area section	A_T = 535.27 cm² ; α = 2.527
mass flow ratio $\dot{m}(LOX)/\dot{m}(LH)$ = 5.89	
steady total mass flow rate	$\dot{m}$ = 232.25 kg/s
specific reference enthalpy	h_F = - 938.5 J/g

<u>Table 1.2</u>: Basis parameters for a HM 60/1 rocket engine

The NASA-Lewis Code presents, with this data, a printed form in which
numerous combustion gas quantities are tabulated as functions of the
area cross section ratio ϵ of nozzle exit to nozzle throat. For this
well-known form sheet the equilibrium properties of state are calculated for a chosen steady mass flow rate of the gases according to the AFC
method explained in Section (1.2).

In the following discussion the value ϵ = 45 was selected for the HM
60/1-engine as reference area ratio. All relevant gas flow properties
in the nozzle exit section are cited in <u>Table 1.3</u>.

Results are obtained with thermodynamic data for the species of the
burning gas[34], compiled in the Lewis computer equilibrium program.

properties of state in the nozzle exit section			
pressure	p	[bar]	0.1808
temperature	T	[K]	1356
specific enthalpy	h	[J/g]	-10124
mean molar mass	ψ	[g/mol]	13.89
specific entropy	s	[J/g·K]	17.78
mass density	ρ	[g/m³]	22.27
Mach number	M	[-]	4.26
flow velocity	v	[m/s]	4286
characteristic velocity	c^*	[m/s]	2328
(reactive) isentropic exponent	κ_s	[-]	1.247
mol fraction H_2	x_{H_2}	[-]	0.2579
mol fraction H_2O	x_{H_2O}	[-]	0.7421
theoretical vacuum impulse	I_{sp}^{vak}	[s]	456.3

Table 1.3: Compilation of HM 60/1 values for ϵ = 45

For the calculation of the properties of state in the nozzle exit, S. Gordon & B.J. McBride assume that there are not only the relationships from the section (1.2), but also all values in $\underline{E}$ are dependent parametrically on the values in $\underline{I}$ and explicitly on the area ratio $\epsilon := A_E/A_T$. With these assumptions, they introduce an iteration procedure which is as physically unjustifiable as Prozan's hypothesis: since the local pressure in the expansion section of the nozzle usually drops sharply in comparison to the local temperature, they postulate a set of empirical formulas to relate p_E to ϵ ; an example is previously given with equation (1.34).

In order to obtain pressure p_E , the NASA-Lewis Code recommends (p.40) calculating it from the correction formula

$$(1.43) \qquad (p_E)_{j+1} = (p_E)_j \left[\frac{\epsilon_{j=0} + \Delta\epsilon_j}{\epsilon} \right]^{\alpha_j}$$

with the help of a **sequence** of $\Delta\epsilon_j$-values (j = 0(1)10); the exponent α_j is, through the empirical relationship (115) in the NASA-Lewis Code, related as parameter to other, considerably less variable properties of state in **E**. Additionally, the (reactive) isentropic exponent κ_s determined for throat conditions affects α_j. For j = 0, equation (1.34) can be used to determine the initial value $(p_E)_{j=0}$.

Each element in the sequence $\epsilon_{j+1} := \epsilon_j + \Delta\epsilon_j$; $j = 0(1)J^{35}$ corresponds, through the energy equation, continuity equation or thermal equation of state, to a set of properties of state $(p_E)_{j+1}$, $(T_E)_{j+1}$, $(v_E)_{j+1}$ and $(\rho_E)_{j+1}$, which are assigned to the mole fractions $(X_{k,E})_j$ according to the AFC method. The sequence converges, whereby the convergence is guaranteed through the fulfilling of the isentropic condition (1.32) within a given limit for the criterion $\epsilon_j \approx \epsilon := A_E/A_T$.

<u>Table 1.4</u> lists pressures p_E in the nozzle outlet section (from MBB calculations for the ARIANE-HM60/1 engine) for some of the given values of the area ratio ϵ. These values are compared with the initial values $\epsilon_{j=0}$, for which 'convergence' can be obtained at J ≤ 10.

$\epsilon = A_E/A_T$	10	15	30	45	60
p_E[bar]	1.33	0.776	0.308	0.180	0.115
$\epsilon_{j=0}$(J≤10)	10.007	14.963	29.77	44.52	59.55
$(\approx -10^2 \Delta\epsilon_j/\epsilon)$[%]	-10^{-2}	1/4	0.8	1.2	3/4

<u>Table 1.4</u>: Initial values for the Lewis Code iteration with prescribed sequence $p_E(\epsilon)$ - 'matching 1' -

In another line of the table one can see at which distance (in %) from the reference value ϵ one has to start with the iteration, in order to meet the final value p_E precisely with 10 iteration steps: a systematics is not discernible! Of course the question asked is usually reversed: at a given area ratio ϵ and known parameters in $\underline{I}$, how high is the pressure p_E, with isentropic expansion and chemical equilibrium?

In order to demonstrate the extent to which the calculated value of p_E depends on the details of the described iteration procedure, all values of the properties listed in Table 1.3 were established in the following way: a given number of iteration steps in relation to the area ratio ϵ = 45 was used for each of the completely different initial values $\epsilon_{j=o}$ ('matching 2'). When the other parameters are not altered (see Table 1.2) and the isentropic condition is fulfilled, a discrepancy of up to 60% (relative to the largest occurring p_E value) in pressure and up to 30% in vacuum impulse is noted, depending on the initial value chosen for the iteration.

Thus it is not surprising when one, using a modification of Table 1.4 ('matching 3'), arrives at highly differing final pressures p_E , along with significant variations in the corresponding $I_{sp,E}^{vac}$-values, for some fixed area ratios ϵ but variable initial values $\epsilon_{j=o}$.

If one does not want to reject this procedure, these uncertainties can only be eliminated with supplementary and independent information, assuming that one can use the NASA-Lewis Code approximatively also for the simulation of real processes. Such an application explicitly contradicts the stipulated premises regarding dissipation-free processes in the combustion chamber/Laval nozzle configuration.

In this questionable case, one wants to equate the experimentally (under the same operating conditions) determined $I_{sp,E}^{vac}$-value with the theoretical value, in order to avoid basing a decision about 'correct' theoretical results on a random initial value or a non-verifiable limited number of iterations. If this is done, the corresponding final pressure p_E is also known – a value which is a purely theoretical information – solely relevant in relation to other consistent values of the theoretical model chosen.

In <u>Table 1.5</u> **wall pressure measurements** compiled for some known rocket engines may be compared[36] with corresponding p_E values. PIRUMOV & ROS-LYAKOV (1986, p.348) offer interesting information on the theoretical background of such pressure distributions along the nozzle radius.

* rocket engines		J-2	SSME	HM7	HM60	HM60/I
chamber pressure	p_F[bar]	54	205	35	100	100
nozzle area ratio ϵ		27.5	77.5	83	106	37.5
nozzle exit pressure						
– one-dimensional	p_E[bar]	0.189	0.180	0.029	0.061	0.236
nozzle exit pressure						
– wall –	p_E^W[bar]	0.265	0.337	0.050	0.129	0.364

<u>Table 1.5</u>: Nozzle exit pressures for selected rocket engines (p_E^W-values compiled from Messerschmidt-Bölkow-Blohm, Munich)

Such banal matching of physically untenable theoretical and experimental information is common practice. It seems that a computer program which hasn't been revised for over twenty years and was used so utterly uncritically apparently assumes a sacrosanct status in the minds of its users: matching may lead to 'truth`, and combustion efficiency becomes one (or even above)!

1.6 General Commentary

It should be emphasized at this point that this study's analysis of the investigated NASA codes with their often significant defects already provides an important result sui generis. Everyone who uses the NASA-Lewis Code and who is informed about these defects must be concerned and take them seriously. Such critical reservations currently prevail at NASA's Marshall Space Flight Center. At any rate, attempts have been made there since 1985 to produce an efficient and valid modification of the NASA-Lewis Code. This fact should be noted by all those readers who

may be, for whatever reason, skeptical of the problem's solution – the
Munich Method – which follows. At least they are informed and will be
spared a retreat to the status quo ante.

While honoring the world-wide positive experiences with the NASA-Lewis
Code, the significant result of the following analysis must be put on
record: with considerable experience (above all, with the ϵ iterations)
and using the AFC method for contraction values $\alpha > 2.5$, one is apparently able to obtain information about the equilibrium properties of
state and the specific vacuum impulse in the nozzle outlet section that
corresponds with experience (in the H_2-O_2 combustion). These results
demonstrate the high quality of Gordon & McBride's equilibrium code and
the insensibility of the specific impulse to computational variances.

This method, however, fails to provide an optimal solution. Due to proven theoretical flaws in the section used to calculate performance, the
method is neither recommendable as a planning instrumentarium for the
systematic variation of all relevant operating parameters nor suitable
as a basis for the development of new long-range modes of calculation.
In other words: **the greatest flaw in the NASA-Lewis Code and its variations is without a doubt the unreliability of its basic rocket performance theory.** As a method of approximation it is scarcely calculable.
As a standard for a comparative procedure used, for example, to calculate regenerative cooling and hypersonic combustion processes, the Code
is ill conditioned, particularly when one considers the expense of such
developmental programs. An extension of the code must focus on the performance calculations rather than on the theory and computational program for complex chemical equilibria.

The functions of required thermodynamic data of all relevant components
during combustion and their effects on computer calculations are not
discussed in current literature.

"...Thus there is a flow mechanics for liquids, but none for gases."

-G. Falk-

2. Basic Principles of the Alternative Theory

2.1 Introduction and Reference to the Microphysical Fundamentals

In order to solve the problem of finding an inherently consistent formulation of a provable ideal thermodynamic comparative process for
evaluating relaxing flows (ICP), extensive theoretical work is unavoidable. This is due less to the need for an appropriate mathematical apparatus to formulate the ICP than to the fact that current paradigmati-
cally-oriented physical theories simply do not permit the type of solution sought here. This latter circumstance is not readily apparent in
standard texts: the problem of mapping theories on processes of microphysical systems (of mechanical-statistical or kinetic nature) to pertinent macroscopic theories[37] appears to be either of no interest or
definitively solved. Yet is it actually possible, for example, to derive the Navier-Stokes equations from the Boltzmann equation or vice
versa?

The primary problem in using the Boltzmann equation lies in its possibilities for mapping real micro-processes to the level of pertinent
macro-phenomena connected with a hierarchy of macroscopic equation
sets. Thus it is imperative to establish clear relationships between
the Boltzmann equation and each stage in this hierarchy. Yet the reasoning for the hierarchy presented in given texts is highly controversial. Alone in the popular discipline of 'irreversible thermodynamics'
there are at least three different 'schools' currently accusing one
other of heresy. These circumstances must be taken into consideration
before one can even begin to understand and treat the problem at hand.

A concise presentation here, however, requires more space than is available in this study. Extensive investigations of this problem have already been made (STRAUB 1986, 1987; STRAUB & LIPPIG 1976). In a supplement, J. Keller (1987) offers a short comparison of the theories of irreversible processes giving the reader quick and relatively unprejudiced access to the most important sources.

The author's investigations of the microphysical fundamentals of thermofluiddynamics were preceded or accompanied by a series of studies on gas kinetics (see the bibliography in STRAUB 1987). Whereas they deal with theoretically fundamental or practically relevant problems of the transfer mechanisms in gases and fluid mixtures, the investigations mentioned above were based on the work done by I. Prigogine and his associates in the past decade.

The results of all these studies focus on the elementary role of irreversibility in the micro-, as well as macrophysical world. They can be summarized briefly as follows:

(1) Insofar as macroscopic physical systems can be mapped on the microphysical level using a multiple particle system with a continuous spectrum of energy eigenvalues, the following facts must be considered with regard to thermodynamics:

   (a) Dissipation is a fundamental effect in the **micro**physical world, and divides the real dynamics of the elementary particles into reversible and irreversible processes.
   (b) One can construct a Ljapunov functional as a connection between micro- and macrophysical entropies (see LAVENDA 1985, p.60).
   (c) Prerequisite for (a) and (b) is the existence of a **time operator** of the particle system with non-unitary transformational properties. This time operator also allows a non-contradictory introduction of energy eigenvalues.

(2) The macroscopic laws, especially the balance equations and constitutive functionals, must be thoroughly revised according to Prigogine's results. Justifying them with the Boltzmann equation, as TRUESDELL (1984, pp.413 f) does, is untenable!

This is also true for quantum theory as a special case: a significant contribution, as G. Ludwig (in MEHRA (Ed.) 1973, pp.702 f) demanded, is obtained for its substantiation!

The most peculiar result is, without doubt, the refutation of the **principle of microscopic reversibility**[38] (see FALK & RUPPEL 1983, p.106). Another fully unexpected result is the establishment of natural times as a concrete manifestation of a microscopic time operator. Its existence is primarily guaranteed by the fact that, with its help, the central particle number term for all macroscopic theories can be uniquely defined (see STRAUB 1987, pp.231 f) – when one has to abandon the classical trajectory concept on the micro level. The molecular trajectories of a gas can no longer even be "calculated in principle" (see ULMER et al. 1987, p.149) and a hard core model, for instance, is useful only for primitive concepts in kinetics. D'ESPAGNAT's assertion (1971, p.7) that "the particles are themselves 'physical properties'" is thus confirmed.

Proof of a Ljapunov functional for the multi-particle system mentioned above offers a second important connection between the macro and micro levels. Pertinent consequences are also drawn for the relationship between time operator and the usual macroscopic time as process parameter, and also affect equilibrium and stability problems of open systems, according to the Second Law of thermodynamics. The following circumstances are of particular interest: although a close abstract link between the micro- and macroscopic levels of a multi-particle system can be constructed, it is basically impossible to establish precisely and in view of a concrete microsystem the pertinent relations between the properties of a corresponding macrosystem. The reason lies primarily in the **non-linear** structure of macroscopic balance equations. It is possible, however, to formulate a correct H-theorem for important model systems – contrary to Boltzmann's famous result (see STRAUB 1987, pp. 240 f).

Generally speaking, Prigogine's dynamics provides not only an inherently non-contradictory microscopic theory for establishing the Second Law of thermodynamics on the atomic level, but also valuable instructions for developing a deductive macroscopic continuum theory for real fluids. Since this dynamics is not, unlike common statistical thermodynamics, mainly limited to states of equilibrium, it is possible to construct a suitable Ljapunov functional $\Omega^{micro}(\ldots|t)$ for a microscopic multi-particle system. With this basis, at least, one can justify the hypothesis of using such a functional $\Omega^{macro}(\ldots|t)$ for describing macroscopic non-equilibrium processes. KREUZER (1983, pp.235 f) has discussed its premises at some length. The decisive argument used here is the analogous possibility of defining $\Omega^{macro}(\ldots|t)$; it involves a description of thermodynamic stability using the entropy of the system (see STRAUB 1987, pp.178 f). It should be noted that an entropy concept is used for representing the non-equilibrium processes of an open system, insofar as it is possible to be introduced as an appropriate physical property describing complex processes.

A comment made in Ulmer's book by a scientist in a dialogue with a philosopher offers a fitting summary: "I believe I understand – Newton's concepts of space and time, for example, are still valid, yet only within certain limits. They are still operable in mechanics, but since they are true only within limits, their meanings change." (ULMER et al. 1987, p.156).
In addition, I. Progogine quotes A. Koyré: "The legacy of Newtonian mechanics is a concept of a world in which **time is essentially a parameter associated with motion**. ... relativity and quantum mechanics have kept unmodified this character of Newton's world model." (see in MEHRA (Ed.) 1973, p.561).

2.2 Forms of Energy and Gibbs Fundamental Equation

It was a fortunate coincidence that at the same time Prigogine's Brussels school presented its revolutionary ideas on a new microscopic theory of matter, G. Falk was reworking his strictly formal axiomatics of thermodynamics into a rather abstract and innovative theory of macroscopic physics. The latter was subsequently elaborated and published, together with W. Ruppel, in two volumes as a transparent and virtually universal method for describing physical processes of multi-particle systems. The rather worn term "dynamics", also used by Falk, apparently represents a program: although he shares, for example, H. Hertz's sharp criticism of the traditional idea of force, he decidedly opposes the assertion that "kinematics, as the intrinsic connection between geometry and time, have a logical and methodological priority over dynamics. Reducing terminology of dynamics on kinematics was considered the primary task of theoretical mechanics." (JAMMER 1981, p.95).

Falk repeatedly points out that (physical or processual) time is not a dynamic variable.[39]According to him, the only such variables are those which define a so-called **Gibbs function** as synonym for the definition of the physical system. At this point, one should refer to the extensive proofs and carefully chosen examples that make Falk's dynamics so attractive (see FALK 1966 and FALK & RUPPEL 1976/83).

In order to offer a compact presentation, particular items are confined to the relevant terminology and the functional connections between the main properties. Appropriate references are also quoted. Additional comments are limited to key informations relevant to the symbols used and to an explanation of the fundamental assumptions for the theory.

The fundamental terms in Falk's "General Dynamics" (FALK 1968, p.1) are **'state'** and **'variable'**, each of which has a specific value for each state.

Another fundamental concept is the term of the (thermodynamic) **'system'** which can be explained abstractly as the total of all states. It is possible, however, to express every (thermofluiddynamic) system with the set of its **dynamic standard variables**. Nature can thus be described using variables, regardless of whether they refer to mechanical, electrical, optical, chemical, or thermal processes. Essential for such a mapping are the energy forms EF_i, each marked by the index i and defined by the equation

$$(2.1) \qquad\qquad EF_i := \varsigma_i \, dZ_i \quad .$$

The mathematical structure for an **energy form** is identical in all fields of physics. Rational representations of physical occurrences, in Ernst Mach's sense, are thus possible. The various fields of physics involved here are not individualized, thus allowing the inclusion of quantum phenomena in the theory. Each member of the energy forms also permits an uncomplicated interpretation of the **'principle of material**

objectivity`, since its mathematical structure and physical meaning are completely independent of the system's observer.

The broad generality of this conceptual apparatus leads Falk to observe: "Thermodynamics includes the mechanical processes. In addition, it also encompasses all other physical processes." Further: "Thermodynamics is restricted neither to heat nor to any special objects or processes in nature, as mechanics is bound to the mechanical, optics to the optical and electromechanics to the electromagnetical items. Instead, it deals with rules which are applicable to any number of objects and processes; it is a **general procedure for describing nature.**" (FALK & RUPPEL 1976, pp.152 & V).[40]

Naturally such an all-encompassing statement – which implies that thermodynamical and physical systems are identical – is valid for Falk's 'General Dynamics`, but not for such disciplines presently known as Heat transfer, Technical or Irreversible, or Rational Thermodynamics. Accordingly, the number of dynamic standard variables used in thermodynamics in general is quite large. For practically all relevant scientific/technical problems, however, these variables are easily obtained, since most of them are typified by frequently reoccurring energy forms. Table 2.1 (for phase of a physical system) offers a selection of the most important standard variables Z_i, their conjugated variables ζ_i and notations of their constituent energy forms EF_i (see FALK & RUPPEL 1976 pp.91 & 134).

The 'energy form` is one of the most important fundamentals of Falk's 'General Dynamics`. He states: "Energy forms can only be referred to as **changes** of state. The energy of a system itself cannot be decomposed into forms." (FALK & RUPPEL 1983, p.125).

In general one cannot express the i^{-th} energy form EF_i as a total differential dE_i , even when the conjugated variable ζ_i is supposed to be dependent only on 'its` standard variables Z_i – a special case of integrable systems which still play an important role in theoretical mechanics.

standard variable Z_i	conjugated variable ς_i	energy form $\varsigma_i dZ_i$ of
P momentum vector	**v** velocity	motion
L angular mom.vect.	**Ω** angular velocity	rotational energy
r position vector	**-F** body force vector	displacement
$\underset{\sim}{\sigma}$ Piola-Kirchhoff-stress tensor	$\underset{\sim}{F}$ deformation gradient	shearing
p dipole moment	**E** electric field intensity	dielectric polarization
m magnetization	**M** magnetic intensity	paramagnetic solids
Q electric charge	ϕ electric potential	electromag. field
A interface vector	σ surface tension tensor	interfacial tension
V volume	-p pressure	compression
N particle number	μ^t chemical potential	chemical energy
N particle number	η^t electrochem.potential	electrochem.energy
S entropy	T temperature	system configuration

Table 2.1: The most important standard variables and energy forms

Some questions arise: why are the energy forms listed this way and not formed some other way, i.e.: why is the energy form of motion, for example, called **v**·d**P** and not (as known from point mechanics) **P**·d**v**? And why don't terms such as energy, heat, work, kinetic energy etc. appear in the list?[41]

The answer is fundamental for Falk's 'General Dynamics':

(1) Every physical system has energy E.

(2) The identity and inequality of systems can be seen from the respective relationships between E and its attributed variables. Of particular fundamental importance is the relationship between E and the standard variables of the involved energy forms: this relation-

ship is called the **Gibbs-Euler Function (GEF)**

$$(2.2) \qquad E = \hat{E}(\ \mathbf{P}, \ \mathbf{r}, \ \mathbf{L}, \ \underline{\sigma} \ , \ \mathbf{p}, \ \mathbf{m}, \ Q, \ \mathbf{A}, \ S, \ V, \ N, \ldots)$$

(3) The GEF of a system contains all physically relevant informations; both definitions are synonymous and can be used interchangeably. The GEF is the most important of all thermodynamic potentials (or Gibbs functions), which are formally considered equivalent since they also include all system information. Their variables are in part not the variables of the standard energy forms, thus the energy E is also not the Gibbs function of the system.

Of practical importance is the possibility that all thermodynamic potentials, including the GEF, are interconvertible, using **Legendre transformations** as a matter of simple convenience.

One can thus transform the Gibbs Euler Function E with

$$(2.3) \qquad E_{LT} = E - \sum_{\ell=\ell'}^{L} \frac{\partial E}{\partial Z_\ell} \ Z_\ell$$

to the Gibbs function $E_{LT} = \hat{E}_{LT}(Z_{i \neq \ell}; \ \partial E / \partial Z_\ell)$, $i=1(1)\ell'..L..I$.

"This is, admittedly, a convenience without which thermodynamics would be almost unusably awkward, but in principle it is still only a luxury rather than a logical necessity." (CALLEN 1966, p.91).

(4) The GEF, as thermodynamic potential for the system's energy E, is 'exclusive` because it depends on variables Z_i , which all have eight (!) properties in common or are allocated to a field (for example, **r**); the variables E, $Z_1, \ldots, Z_I$ are called **extensive**[42] , and the energy conjugated ς_i are denoted as **intensive** variables. [*)]

[*)] Traditionally oriented textbooks frequently offer a confined or wrong interpretation of these important terms : "Thus an intensive property may be formed from any extensive property through division by any other extensive property." (WILLIAMS 1985, p.522).

Every extensive variable has the following attributes, regardless
of whether they are relevant to a system or not:

1. it is proportional to particle number	5. it is defined also for non-equilibria
2. it has a 'density' Z_V	6. it has a fixed zero point
3. it satisfies a conservation theorem	7. it realizes processes
4. it is assigned to a 'current density' j_Z	8. it establishes one energy form

Table 2.2: Attributes of every extensive variable

These facts are significant for physics: they are the results of
over three hundred years of collective experience gradually culmin-
ating in a solution for identifying a set of universal variables
with an astoundingly large number of identical qualities. **This cu-
mulative experience is the empirical-heuristic basis of Falk's
"General Dynamics".**

The physical significance of each individual attribute is explained
in the following sections. It should be noted here that attributes
5 to 8 are also wholly or partially applicable to intensive variables or other properties (heat, for example), yet not to attributes 1 to 4! Degenerative cases are also included: volume V, for
example, has 'volume density one'. Similarly, it suffices if variables satisfy the pertinent conservation law for only particular
classes of processes[43]. In these special cases the conservation
theorems are either natural laws or adjusted to concrete problems
(for example, V = constant).

Extensive variables are the natural exchange properties: each extensive variable defines an energy form in such a manner that the
expression ςdZ always implies an unequivocally assigned intensive
variable ς . Of course all energy forms need not be dependent on
one another. The number of a system's independent variables corresponds to its **degrees of freedom.**

Attribute 5 is frequently misunderstood: since variables V or S are extensive, suitable for describing composed systems and thus physically adequate for non-equilibrium states, their conjugated properties of state -p and T attain physical significance with their corresponding energy forms. Naturally these intensive variables are by no means — as it is often still required in texts today — defined with appropriate 'methods of measurements': these naive 'operational' concepts are criticized by FALK (1968, pp.2 f) as well as by TRUESDELL (1984, p.27). In complex non-equilibrium processes, experimental checks of theoretical investigations are generally only indirectly possible. That is, certain characteristic bulk properties (such as the thrust of a rocket engine) can be determined theoretically as well as experimentally: the degree of coincidence allows reliable conclusions as to whether the chosen 'non-equilibrium equations' are sufficient for the problem or not.

(5) While energy E of every system is determined by its GEF, terms such as **heat** and **work** do not immediately concern the system itself, but rather refer to interactions between the system and its 'surroundings'. This point will be treated in greater detail later. The concept of kinetic energy in the Alternative Theory can no longer be naively interpreted: it will be discussed in the following chapter.

Due to its immediate relationship to these irreversible interactions (expressed by the First Law of thermodynamics), energy is distinguished from other system variables such as entropy S: as a result Gibbs functions like $\hat{E}(...S...)$ and $\hat{S}(...E...)$ are not completely equivalent.

According to Falk: "The system equals every formal pattern whose extensive variables indicate the relationships given by energy as Gibbs function... one can regard the Gibbs function as the system better than any such pattern." (FALK & RUPPEL 1976, p.134). The GEF $E = \hat{E}$ (all independent extensive variables of the system) thus contains the complete physical information on the observed system. **The system itself is always a model of a real concrete slice of nature. It is not, however, a copy of the real original, but rather an abstract model, a mathematical**

construction of a finite number of variables usually ignoring one or more less important aspects of the real situation.

An essential element of every physical system is the exchange of energy in specific forms. All exchange processes in a particular system are represented by appropriate energy forms. The total **energy changes** induced by the exchange processes dE of the system are simply the sum of the participating energy forms

$$(2.4) \qquad dE = \sum_i \zeta_i dZ_i \quad .$$

To allow integration of the linear **Pfaffian differential forms** $\zeta_i dZ_i$, it must be agreed that equation (2.4) contains **all** independent energy forms which can exchange energy in any way. Under these conditions, dE becomes the total differential of the GEF. It is, by definition, always satisfied if the system is an adequately realistic model of the reality to be mapped.

The total differential of the GEF – equation (2.2) –

$$(2.5) \qquad dE = \mathbf{v} \cdot d\mathbf{P} - \mathbf{F} \cdot d\mathbf{r} + T_* \, dS - p_* \, dV + \sum_k^K \mu_{k*}^t dN_k$$

makes all energy forms not participating in exchange processes, i.e. system changes, (such as here, for instant, the extensive variables L, $\underline{\sigma}$, p, m, Q, and A) equal to zero. One can achieve this by keeping the affected extensive variables constant, or by maintaining the conjugated variables at a zero value. The energy form $T_* dS$ has an exceptional position: it occurs practically in every process with real materials and thus is a fundamental contribution to the system's configuration.

Every total differential of the GEF, such as equation (2.5), is called the **Gibbs Fundamental equation** (GFE) of the system. It is a fundamental of Falk's "General Dynamics" and represents the empirical-heurististic basis of this theory. The asterik next to T, p, and μ_k indicates that the GFE in the form of equation (2.5) contains the Hamilton energy forms $\mathbf{v} \cdot d\mathbf{P}$ and $\mathbf{F} \cdot d\mathbf{r}$ – and thus signalizes a moving system. By the way, one can obtain the intensive variables directly from the GFE using the appropriate differentiation requirements[44]. The following equations for the system's **temperature** and **pressure**, for example, are immediately derived from equation (2.5)

(2.6)
$$T_* = \frac{\partial}{\partial S} \hat{E}(P, r, S, V, N_k) := \left[\frac{\partial E}{\partial S}\right]_{P,r,V,N_k}$$

and

(2.7)
$$-p_* = \frac{\partial}{\partial V} \hat{E}(P, r, S, V, N_k) := \left[\frac{\partial E}{\partial V}\right]_{P,r,S,N_k}$$

An interesting aspect is revealed by both these equations and deserves closer attention: the multiple **interdependence** between the individual energy forms! They simply result from the dependence of E on a set of relevant extensive variables. Thus in determining the properties of state with equations such as (2.6) or (2.7), the set variables used in the differentiation provide important information about the system of interacting variables, regardless which of the two formulations are used. In this context, the special case of the **state of rest**

(2.8)
$$P := 0 \rightarrow \begin{bmatrix} v = 0 \\ \Omega = 0 \\ r = r_0 \end{bmatrix}$$

is of particular theoretical and practical significance. The above equations for the temperature and pressure of the system, for example, are altered in a very characteristic way:

(2.9)
$$T = \frac{\partial}{\partial S} \hat{E}(P=0, r=r_0, S, V, N_k) := \left[\frac{\partial E}{\partial S}\right]_{P=0,r=r_0,V,N_k}$$

(2.10)
$$-p = \frac{\partial}{\partial V} \hat{E}(P=0, r=r_0, S, V, N_k) := \left[\frac{\partial E}{\partial V}\right]_{P=0,r=r_0,S,N_k}$$

In this case the state of rest is defined as a state of equilibrium. This is not true for cases $v = 0$.

2.3 Gibbs-Duhem Equation, Process and Realization

The most important conclusion is immediately derived from the existence
of the GEF – equation (2.2) – or, more precisely, from the extensivity
of all variables Z_i and their elementary association with the pertinent
energy forms. The GEF is a **homogeneous first-order function** in Z_i. Using the well-known Euler theorem for such functions (see CALLEN 1966,
pp.26 & 47), one gets the general expression

$$(2.11) \qquad E = E_\# + \sum_i \varsigma_i \, Z_i$$

for the system's energy. For systems with standard variables according
to the GFE (2.5), equation (2.11) is specialized to

$$(2.12) \qquad E = E_\# + \mathbf{v} \cdot \mathbf{P} - \mathbf{F} \cdot \mathbf{r} + T_* S - p_* V + \sum_k \mu_{k*}^t N_k \; .$$

$E_\#$ stands for the rest energy (zero point energy). It corresponds to
the rest mass m of the system, $E_\# = mc^2$, in which c means the speed of
light. At this point it is essential to know which N_k particles are active in the system's energy exchange processes. For macroscopic physics
and all engineering sciences (with the possible exception of nuclear
engineering) only selected particle populations can be considered:
those which are governed by the **conservation law for baryons** (see
STRAUB 1987, p.21) during the exchange. The thermodynamic concept of
mass m_B is thus related to the baryon particles (= protons, neutrons,
hyperons and their associated antiparticles – see, for example, FALK
1966, pp.112/3) which constitute virtually all atoms, molecules, ions
etc. in chemical reactions. Using the formula $m = m_B + \Delta m$, one can introduce the **specific** variables through division $z_i := Z_i/m_B$ instead of
the extensive properties of state; the expression

$$(2.13) \qquad e = \mathbf{v} \cdot \mathbf{i} - \mathbf{f} \cdot \mathbf{r} + T_* s - p_* \rho^{-1} + \sum_{k=1}^{K} \mu_{k*} \omega_k + (1 + \frac{\Delta m}{m_B}) \, c^2$$

is obtained for the system's Gibbs-Euler Function (GEF). Compared with
the GEF of equation (2.12), there are new properties of state: the specific energy e, the specific impulse $\mathbf{i}$, the specific field force $\mathbf{f}$, the
specific entropy s, the mass density ρ, the chemical potential per mass

unit μ_{k*} and the mass fraction ω_k of the k^{-th} kind of the mixture of K components.[45]

If a total differential (see the section above) is constructed for property e using equation (2.13), relationship

(2.14)
$$\sum_k \omega_k \, d\mu_{k*} = \rho^{-1} \, dp_* - s \, dT_* - \mathbf{i} \cdot d\mathbf{v} + \mathbf{r} \cdot d\mathbf{f} \quad ; \quad k = 1(1)K \quad ;$$

results from the comparison with the GFE – equation (2.5) – divided by the unchanged baryon mass m_B: this is then known as the **Gibbs-Duhem Equation** (GDE). If it is simplified for a uniform material (K = 1) and the state of rest ($\mathbf{i} = 0$, $\mathbf{r} = \mathbf{r}_0 \equiv 0$) is used as a reference point, one arrives at the expression

(2.15)
$$d\mu = \rho^{-1} \, dp - s \, dT \quad ;$$

this states that the chemical potential μ of the material is a function of T and p.

The theoretical significance of the GDE can be summed up as follows: it promotes the **existence of equations of state** – for example, in the unusual form $\mu_{k*} = \mu_{k*}(\omega_j, p_*, T_*, \mathbf{v}, \mathbf{f})$ for each component k or j; (k,j = 1(1)K) – as a necessary condition for the empirical-heuristic basis of Falk's "General Dynamics", as expressed in the GEF as well as the GFE.

From a phenomenological standpoint, these equations of state can only be determined experimentally or through auxillary theories taken from statistical thermodynamics. Since this is a significant problem, K. Lucas' recently published text "Applied Statistical Thermodynamics" is recommended. The significant difference between the terms 'process' and 'realization of process', particularly with regard to practical applications, are made especially clear in the book.

A **process** is a series of infinitesimal transitions or a **sequence of states**, since every transition has its own initial and final state. Each **realization of transitions or processes** of a system, on the other hand, is the **operational organization of constraints** under which certain processes can or cannot take place. Realization in particular means a defined interaction between system and 'environment'. The system's changes of variables are correlated with the changes of variables of other systems. The isolation of a system against the exchange of certain properties is also such an interaction.

It is now important that the GFE of each system relates only to the processes, not their realization. The same holds for the Gibbs-Duhem equation (2.14). One can thus integrate this equation between any given initial state and a chosen final state. In doing so, the technique of integration can be set so that the mathematical integration procedure is easily carried out. In traditional thermodynamics, this method is routinely used for calculating the equations of state for real fluids. For the chemical potential μ_k, for example, one finds the solution in, e.g., PRAUSNITZ 1969, pp.30 & 41. This solution assumes an integration path for equation (2.14), along which the system is at rest ($i = 0$), supposing an unchanged specific field force ($df = 0$). Each chemical potential μ_k is thus known to be dependent only on the mass fraction of one or more components as well as on the **pressure and temperature at rest**, according to equations (2.9) and (2.10). If one wishes to realize certain processes in a moving system, this limited information at least offers important clues for the further development of the theory.

Technical applications are governed by the realization of processes. Since processes dominate the engineering sciences, a few of their salient aspects should be briefly mentioned here to prevent subsequent misunderstandings (see FALK & RUPPEL 1976, p.257). The following points are significant:

(i)   the relationship between the terms 'state' and 'time',

(ii)   the general significance of the physical definition of equilibrium,

(iii) reversible and irreversible process realizations.

With regard to (i), it should be mentioned here that "although a physical system always is in a definite state at a given moment, it cannot be said that the property 'time' has a given value in a specified state. **Time** thus is not one of the **dynamic properties**, and therefore cannot appear as an independent variable in the Gibbs functions. The propagation of time manifests itself by the system's run through a sequence of states, or undergoes a process." (FALK & RUPPEL 1976, p.148). The manner in which this process is realized depends on the **conditions of realization**. Of particular importance, for example, is the **adiabatic process realization**: it is by no means tied to Carathéodory's 'adiabatic wall`, but can also be realized through 'time` - i.e., through a relatively 'fast` process of changing states! In this case, 'time` marks the specific value of a suitable curve parameter.

Similarly with the definition of equilibrium: "It is important to comprehend that the description of the equilibrium does not depend on a sequence of time; it is established only by the **quality of an equilibrium state**, in so far as this does not require the possibility of a free exchange of energy forms." (FALK & RUPPEL 1976, p.196).
"Equilibrium thermodynamics does not provide the entropy with a functional dependence on time." (LAVENDA 1985, p.61). L.C. Woods emphasizes: "Equilibrium processes are usually described as being 'quasi-static`, but this is a misleading term, often being misinterpreted as meaning 'static`." (WOODS 1975, p.11).

It is particularly important for the dynamics of a system that its variables, while being affected by other systems, are held constant with realizations by any equilibria. This possibility often leads to an **internal equilibrium of the system**, a term which is virtually synonymous with the concept 'body`: a system whose individual parts are held together through characteristic internal equilibria and which act as a physical unit. It seems that equilibrium is a distribution rather than a state.

This is significant because an internal equilibrium respective to the exchange of an energy form signifies a reduction of the number of independent variables of each relevant Gibbs function. In the GFE, for example, many energy forms of multiple particle systems do not appear. The best-known example is the internal equilibrium between the particles which depict the translatory and rotational molecular degrees of freedom. The findings of L.C. WOODS (1975, p.69) are worth mentioning:

if those energy forms in the GFE which represent relaxation phenomena by changes of appropriate particle numbers are ignored, only an error of the second order occurs.

2.4 Axioms of the Traditional Continuum Theories: a Commentary

There are essential differences between the AT and currently dominant continuum theories: they are detailed in an internal report (STRAUB 1988). Roughly summarized, these **traditional theories** can be character-ized by five theses:

(i)   They are mechanical theories: mechanical principles and the **in-compressibility** of the model fluids are metaphysical premises. Compressibility is simply a Mach number effect; thermodynamics are reduced to the appropriate processing of 'real gas effects'. Theoretical questions of measurements are at best treated as iso-lated mechanical problems. The measurement of pressure is a typi-cal example!

(ii) They are mathematically oriented: the **kinematics** of the processes are emphasized. There are a considerable number of paradox find-ings: so much the worse for the facts! Important theoretical for-mulations concentrating on investigating the dynamics of flows, such as Prandtl's boundary-layer theory, are rarely even mention-ed by most 'schools' (see Section 2.1).

(iii) With the exception of certain formal ties, there is no link be-tween thermodynamic principles and the transport coefficients. In order to obtain equations of motion for 'frictionless' flows, the viscosity is set at zero. **Irreversibility** and dissipation are rarely properly discussed as physical fundamentals, let alone in-tegrated theoretically as an elementary concept in the continuum theory. Viscosity is trivialized to a coefficient in an expansion rule of tensor algebra.

(iv) The problem of **turbulence** is treated either with pure mathema-tics, probability theory, Reynolds methods etc., or completely ignored. Most 'schools' prefer the latter solution! The introduc-tion of Reynolds' random fluctuations ultimately leads to physi-cally abstruse, formal 'models' whose questionable theoretical reasoning, as in the case of the well-known k-ϵ model, is gradually forgotten and ultimately ends up as a meaningless claim.

(v)   All traditional theories are **pseudo-mass-point-theories**: a fundamental element of the continuum is the 'particle', which now is also assigned field properties such as 'temperature'. Nevertheless, Euler's mass point and its laws of motion still are considered as physical fundamentals. Typical of this situation are the close ties with the classical Maxwell-Boltzmann gas kinetics and the Chapman-Enskog interpretation. Modern micro-theories, with their elimination of the trajectory concept and the resulting far-reaching consequences, as yet have absolutely no influence on aerothermodynamical applications.

The historical relationship with mechanics of the 18th and early 19th centuries is evident. That is why the constantly-acclaimed empirical fundamentals of these theories continue to rest on weak premises. H. Poincaré's sober assessment "The principles of mechanics are conventions and disguised definitions" has yet to be refuted. R.J. Mayer's commentary on contemporary science – which "with resignation sets up the axiom that frictional heat is inexplainable" – has been brought up to date with similar explanations in modern texts on flow mechanics (see STRAUB 1986, pp.67 & 87).

One should consider that the majority of the 'laws' of mechanics and flow mechanics still are based on the 'principle of force equilibrium' ('principle of virtual work'). Over one hundred years ago H. Poincaré criticized this basis: "The principle of the equality of action and reaction must not be considered an experimental law but rather a definition." This 'definition', however, disregards the 'interdependence' mentioned above in all concrete applications of the traditional continuum theory. In other words, whether or not the said 'force equilibrium' really exists and, if so, under which constraints, remains completely irrelevant: process realization has nothing to do with traditional flow mechanics.

This fundamental defect is not only of historical nature. It affects, for example, the classical derivative of the Navier-Stokes equation of motion. In 1822, at the time the foundations were being laid for all mathematical versions of our present continuum concepts, neither modern energy nor entropy concepts were known. Far more serious was the inability of scientists and engineers to draw the conclusions from H. Hertz's judgement that the Newtonian definition of force was fully unsuitable for complex models describing nature. This definition of force is factually still paradigmatical![46]This assertion is substantiated by the 'ultramodern' molecular dynamic simulations of sheared flows which have been carried out since 1980 by D.J. Evans & H.J.M. Hanley. Apart from fundamental methodological objections - 'steady', isothermal shear flows with constant shearing for a fluid of **128** Lennard-Jones 'particles' (see HANLEY & EVANS 1982) - the simulation is based on a purely mechanical concept: 'temperature' is simply the kinetic energy of the particle population expressed in Kelvin!

The most important conclusion one can draw from thesis (v) is associated with an inherent existence theorem: at least as far as the motion of one particle is concerned, one always deals with so-called **integrable systems**, which played a significant role in the theoretical mechanics of the 19th and 20th centuries (see GOLDSTEIN 1985, p.337).
As long as one also takes only **conservative force fields** into consideration, the GFE (2.5) can be reduced by the Hamiltonian terms. For the integrable energy forms it is obvious that

$$(2.16.1) \qquad \mathbf{v}\cdot d\mathbf{P} \equiv \mathbf{v}(\mathbf{P})\cdot d\mathbf{P} \quad ; \quad \mathbf{f}\cdot d\mathbf{r} \equiv \mathbf{f}(\mathbf{r})\cdot d\mathbf{r}$$

holds, including the definitions

$$(2.16.2) \qquad \mathbf{v} := \mathbf{P}/m_p \qquad \mathbf{f} := -\nabla e_{pot}$$

for the mass point (with constant mass m_p) in the (scalar) potential field $e_{pot}(\mathbf{r})$. The kinetic and potential energies are eliminated a priori; the **internal energy** of the 'particle' (= system) results! This **a priori identification of real continua with a mathematical mass point model** has led to a concensus in all traditional continuum theories: three model qualities are characteristic for the additional incorporation of important results of **Carnot-Mayer's thermodynamics** into a continuum theory of incompressible fluids:

(1) the 'pressure' occurring in the balance equations can be expressed by the **thermal equation of state**,

(2) the difference between the total energy and the system's **types** of energy – kinetic, potential, electromagnetic energy – is the system's internal energy,

(3) **'external work'** and **heat**, as expressions for the energy transfer between system and **'environment'**, can only be described through system variables.

This consensus is not questioned. It is deeply rooted in the mechanical heat theory and is based on the older, reversible elementary impact axioms for two mass points. With this theory it is possible to construct a formal tie to the early Laplace-Piosson caloric concept and the thermal equation of state in Carnot-Mayer's thermodynamics; assumptions are a few suitable hypotheses for kinetic pressure and temperature. This solution seemed satisfactory: no one bothered investigating the circumstances in which this equation of state for flowing liquids or gases is valid. Since this solution can also be substantiated by methods of statistical thermodynamics, the equilibrium condition needed for this method prompted the introduction of the **'principle of local equilibrium state'** for continua (see VINCENTI & KRUGER 1967, p.237).

In general, contemporary continuum theories exhibit a remarkable convergence: the laws of motion for incompressible mass point model fluids are coupled with the equations of state of compressible fluids at **rest** (= thermostatic); **transfer processes** are treated in the framework of phenomenological and kinetic theories according to the heuristic principles of theoretical mechanics. Although they are frequently interpreted as the manifestation of irreversible processes and thus as an immediate source of local entropy production in the flow, they are as a rule not included in concrete theoretical considerations. Such processes are thus expressed by constitutive 'laws' – e.g. the Fourier- and Fick- 'law' – now as before, for instance, without regard to unsteady states (see STRAUB et al. 1987).

2.5 Significant Results of the Alternative Theory: an Outline

Classic thermodynamics, in J.W. Gibbs' or M. Planck's formulations for heterogenous substances or ideal gases, still offer the most important examples for systems in which the Gibbs function is not separable. Even for the simplest materials, each property of state is dependent on at least two variables. FALK & RUPPEL (1976, p.143) made the additional observation: "For this reason, flow processes in gases, also **cannot** be understood in a purely mechanical sense, unlike flows in liquids, which can be treated – within the approximation presented in view of solids – as a purely mechanical problem. Thus there is a flow mechanics for liquids, but none for gases."

The **Alternative** Theory (**AT**) is just such a flow mechanics for gases. Stated more precisely: the **AT is a phenomenological continuum theory of real compressible fluids**, or a mathematical theory of **thermo-fluid dynamics**.

* Synopsis

The AT is stimulated by Prigogine's new microtheory and involves Falk's dynamics.
Prigogine's particle model is related to his new transformation theory (see MEHRA (Ed.) 1973, pp.561-593) and directly coupled with the system's spectrum of energy eigenvalues.
Prigogine's concept of particles has nothing in common with the individual "aggregates of material points" equipped with 'semispherical diameters' and trajectories which were the basis of the Maxwell-Boltzmann gas kinetics. Nor do they have anything in common, in a more general sense, with "mechanical systems whose conditions are determined by arbitrary generalized (Lagrange) coordinates." [47]

This generalized concept of particles is an important aspect of Falk's dynamics for physical systems. The number of energy forms is also one of its fundamental empirical bases – together with some constraints for transit times for signals involving thermodynamic information.
With the Gibbs-Euler Function as a synomym for the system concept, a **general axiom** is established for an unambiguous definition of states, including those in non-equilibrium. This term is closely identified with the realization of processes: common expressions such as 'quasi-static' or 'local equilibrium' are considered irrelevant.

The concept of field is especially constitutive for Falk's dynamics.
There are two important consequences for a model of the static field
probably representing most non-relativistic macrosystems: first, the
possibility of uniformly establishing a body-field system with its
Gibbs Fundamental Equation; secondly, the formulation of the sought
field equations by utilizing the conservation law characteristic of important system properties mentioned above.

The AT begins on this basis. First, 'particles' are represented as syn-
onyms for simultaneous energy-mass-momentum transports in order to
clarify their essential differences from macroscopic body-field systems.

There is a striking shift in the significance of kinematics for the AT
compared to its role in traditional continuum theories. The flow velocity, for example, becomes primarily dynamic and not kinematic. With
the identification of both definitions an important constituent occurs.

A further constructive condition of general significance concerns the
body-field system: the general axiom is supplemented with some 'principles' of Rational Mechanics: the most important are material objectivity and **equipresence**.
In addition, process time and kinematics are informally introduced in
connection with the First Law of thermodynamics for open systems. This
part of the AT definitions set is particularly controversial since it
is based on a concept of external work that clearly diverges from traditional continuum theories. This concept refers, together with the
heat flow also introduced by the First Law, to the interaction between
the given physical system and its specific environment: it is oriented
toward the customs applied in technical thermodynamics (see STRAUB et
al. 1977).

An especially characteristic result of the AT are the differences found
between the field properties like temperature and pressure, etc. in real flows on the one hand, and the same variables in a hypothetical
state of rest on the other. These differences are the direct result of
the ties between the Gibbs Fundamental Equation for a flowing body-

field system and both the definitions of kinetic energy and the equi-presence principle. A decisive result is obtained from one of this system's typical partial non-linear differential equation: **real body-field systems are not integrable systems; thus local flow velocities and local specific impulses do not identically coincide.** The difference (in a velocity unit) between the two properties of state disappears only in the limiting case of dissipation-free flows: it is thus termed the **dissipation velocity.**

If one restricts the body-field system to a multicomponent single-phase fluid mixture, one may derive the Euler equation of motion with the help of Maxwell's principle of kinetic equilibrium, an explicit formulation of the momentum conservation law, and the conservation of the total energy taken from the general axiom of the AT. This universal method also allows the direct derivation of a Cauchy-type equation of motion, although it has features which deviate considerably from the norm. The flow velocity and pressure in the flow, for example, are clearly defined above the local total energy of the system, and the viscous stress tensor is like a dyadic expression distinguished by the dissipation velocity.

The method presented here indicates the dependence of the dissipation velocity on the square root of the mass density. Thus, for example, all field equations for the local values of velocity, temperature, pressure, density and concentrations of the polynary mixture for the hypothetical state of rest can be formulated. **It must be emphasized that a thermal equation of state can be proven to be constitutive for a flow field only for this special state.** This important theoretical finding presupposes only the general axiom, a few other axioms and definitions, as well as an unambiguous connection between energy and entropy flow vectors, in which the heat flow vector plays a parametric role.

The entropy production density, in its relevant form for property gradients in the flow field, also exhibits a number of features diverging from the norm.

If one further specializes this set of field equations for a polynary, single-phase system as the simplest case of a Newtonian One-component, Single-phase fluid (**Newtonian-OS-fluid**), one observes a number of

striking facts which are especially interesting when compared with the
results of traditional field theories:

(1) the Navier-Stokes equation of motion is, strictly speaking, valid
only for an incompressible model fluid. Together with the incom-
pressibility condition $\mathbf{V} \cdot \mathbf{v} \equiv 0$, it constitutes a closed equation
set for local velocity and flow pressure function. The kinematic
viscosity is a constant: it marks the individuality of the model
fluid.

(2) Real fluids are always compressible fluids. For Newtonian-OS-flu-
ids, motion is determined by the Navier-St.Venant equation of motion. The latter differs considerably from the Navier-Stokes equation of motion: it contains the Euler equation of motion as a limiting case of infinitesimal dissipation. The formal structure of
the expression for the viscous stress tensor also includes a term
which indicates the influence of the entropy flow on the formation
of the velocity field. This expression deserves special attention
with regard to the conventional concept of viscosity.

(3) The first non-trivial approximation of the Navier-St. Venant equation of motion is by no means the Navier-Stokes equation of motion,
but rather a vectoral relationship of the Euler equation of motion
type. It is, however, affected by friction, as can be proven by a
precise preparation of the dissipative flow's pressure in compressible fluid; this result explains the realistic findings recently
obtained in industry using the 'Euler method' even in close proximity to strong compression shocks.

* The Navier-Saint Venant equation of motion

In order to offer a quantitative presentation of the numerous important
results of the AT, the equation of motion of a compressible **multicomponent single-phase fluid mixture** is compared to the Navier-Stokes equation of motion. The basic relationship representing the momentum conservation law is a Cauchy-type equation of motion

$$(2.17) \qquad \rho D\mathbf{v} = \rho f - \mathbf{V} \cdot \overset{\bullet\bullet}{\boldsymbol{\pi}} + \tfrac{1}{2} \, \partial_t \rho \boldsymbol{\phi} \quad ,$$

whereas D is a differential operator called the substantial or material time derivative

(2.18) $\qquad D := \partial_t + \mathbf{v} \cdot \mathbf{\nabla}$;

in this definition ∂_t denotes the partial time derivative and $\mathbf{\nabla}$ is the gradient vector operator (nabla-operator).

The local flow velocity $\mathbf{v}$ involved has a double definition: the original 'dynamic' establishment

(2.19) $\qquad \mathbf{v} = \dfrac{\partial E(\mathbf{P}, \mathbf{r}, s, V, N_k)}{\partial \mathbf{P}}$

is obtained with the Gibbs Fundamental Equation (2.5) of the multicomponent single-phase fluid mixture. Identification of $\mathbf{v}$ with the kinematic velocity $\mathbf{v}_* := d\mathbf{r}/dt$ guarantees conservation of the particle number of all baryons. This invariance generalizes the Newtonian mechanics of finite systems of mass-points, called 'bodies'. At every given instant the body is located at some place and assigned real numbers denoting the properties of state. In field theories the continuity equation controls the time behavior of such a body by considering the baryons constancy (see STRAUB 1988, pp.53 f).

The properties $\overset{\bullet}{\pi}$ and ϕ appearing in equation (2.17) along with mass density ρ and the specific field force $\mathbf{f}$ (for example, gravity), are defined as follows in the AT:

(2.20) $\qquad \overset{\bullet}{\pi} := p\boldsymbol{\delta} - \boldsymbol{\tau}$ $\qquad\qquad$ **pressure tensor**

(2.21) $\qquad \phi := \mathbf{v} - \mathbf{i}$ $\qquad\qquad$ **dissipation velocity**

Here $\boldsymbol{\delta}$ stands for the unity tensor and p is the thermodynamic pressure established with equation (2.10) for the **hypothetical state of rest.** This pressure is permitted to be calculated only with the thermal equation of state

(2.22) $\qquad p = R T \rho Z$

where ρ denotes the density and T is the temperature defined following equation (2.9); the compressibility factor $Z = \hat{Z}(T,\rho)$ is identical to one only with ideal gases,[48] (and along the so-called 'ideal curve').

The viscous stress tensor $\boldsymbol{\tau}$ typical for individual classes of flowing fluids is related to the AT's characteristic dissipation velocity through the dyadic product

(2.23) $\tau := \frac{1}{2} \rho v \phi$ **viscous stress tensor**

The quantity ϕ equals the vector difference between the local flow velocity **v** and the specific (linear) impulse **i**; the limiting case

(2.24) $\lim_{\sigma \to 0} \phi = 0$

disappears only for vanishing dissipation, expressed by the local values of the entropy production density σ.

For $\phi = \hat{\phi}(\sigma \neq 0) \equiv 0$, equation (2.17) formally corresponds to the classical Cauchy equation of motion.

The following **Navier-St.Venant equation of motion** for a defined class of compressible fluids and their mixture is taken from the AT and is comprised of equations (2.17) and (2.20). For the viscous stress tensor τ the expression[49]

(2.25)
$$\tau = t_{\tau} Z^{-1} p \frac{S}{R} \left\{ D - \frac{1}{2}(\nabla \cdot v)\delta \right\} -$$
$$- \frac{1}{2} t_{\tau} T(\partial_t s + \nabla \cdot j_s) \delta -$$
$$- t_{\tau} (ZR)^{-1} p \left\{ \nabla s v - \frac{1}{2}(\nabla \cdot s v)\delta \right\} -$$
$$- t_{\tau} (ZR)^{-1} p \left\{ [\nabla s \times v] \times \delta \right\}$$

is theoretically derived.

For simplification, the abbreviation

(2.26) $t_{\tau} \rho T s := 2\beta = t_{\tau} Z^{-1} p \frac{S}{R}$

is introduced.

Before equation (2.25) is discussed in greater detail, its corresponding traditional continuum mechanics relationship should be offered for comparison. These Navier-Stokes equations, as generalized notations of the set of field equations for $\hat{v}(r,t)$, $\hat{T}(r,t)$, $\hat{p}(r,t)$ and $\hat{\rho}(r,t)$, are **defined** by its viscous stress tensor:

(2.27) $\tau := 2\mu D + (\eta_v - \frac{2}{3} \mu) \delta \nabla \cdot v$.

Perhaps the clearest interpretation of this equation is found in TRUES-
DELL (1984, pp.426-427 & 409-410). In this case, **D** is the deviator
which is the symmetrical share of the (tensorial) velocity gradient **∇v**
in symmetrical stress tensors; μ and η_v are the fluid's **shear or volume
viscosity** respectively. C. Truesdell offers convincing reasons why the
known Stokes relation (i.e. $\eta_v \equiv 0$) is valid.

At first glance one notes that the expression (2.25) for the compressible fluid's viscous stress tensor τ is considerably more complicated
than the formulation – equation (2.27) – for a Navier-Stokes fluid. The
structure of equation (2.25), with its characteristic Nabla operators
(which have multiple effects on the local specific entropy s of the
fluid), contains the asymptotic limiting case

$$(2.28) \qquad \lim_{\sigma \to 0} \tau = 0 \quad ,$$

allowing **a physically satisfactory transition from friction-affected
motion to a dissipation-free Euler flow.** A serious and historically ex-
plicable defect of the Navier-Stokes equation of motion is its lack of
this limit for each infinitesimal local entropy production density σ.
The demand for such an asymptotic behaviour is closely related to Maxwell's principle of **kinetic equilibrium.** On the level of the Maxwell-
Boltzmann gas kinetics, a flow field with gradients results as a solution to the Boltzmann equation when the collision operator disappears
identically for all **r** and t. This case of kinetic equilibrium is asymptotically possible only for reversible processes in which all transfer
current densities equal zero (see TRUESDELL (Ed.), p.414).

Equation (2.25) contains not only one term (which cannot be generally
neglected: see STRAUB 1988, p.123) for the unsteady changes of the specific entropy s. In addition, the divergence of the entropy flow vector
$\mathbf{j}_s$ also appears, which through

$$(2.29) \qquad \mathbf{j}_s = T^{-1} \dot{\mathbf{q}} + \sum_k s_k \, \mathbf{j}_k - (RZ\rho)\mathbf{v}$$

is related not only to the heat flow density $\dot{\mathbf{q}}$ and a convective term,
but to the diffusion flow density $\mathbf{j}_k$ as well. The latter is weighted by
the partial specific entropy s_k of all the polynary single-phase system's components. This divergence term in an expression for a second

order tensor τ apparently contradicts the **Curie principle**. It is a constitutive part of the linear Irreversible Thermodynamics and postulates without proof: even and odd order tensors may not appear together in an equation for continuum mechanics. C. Truesdell vehemently polemicized against this 'principle' (TRUESDELL 1984, pp.387-391). He not only traced its rather dubious origin, but criticized above all its mathematical ambiguity. Equation (2.25) offers an example of the problems encountered with the principle: from divergence of vectors $\dot{q}$ and j_k (k = 1(1)K) follow two scalars. They are tensors of even order zero, whose occurrence together with the viscous stress tensor by no means contradicts the Curie principle. Otherwise the coupling of all transfer processes conforms with all physical experiences; this is not reflected, however, in the Navier-Stokes equation of motion (2.26) (see MOORE 1964, p.192).

This is probably because its 'deduction' fails to take simultaneously occurring transfer processes into consideration. Such a restriction is consequent, however, for an artificial construction such as an **incompressible model fluid**. If one accepts the known incompatibility of this theoretical fluid with the thermodynamics of real materials, one is compelled to acknowledge that concepts such as temperature, entropy, heat flow density etc. are not, strictly speaking, compatible. Taking this circumstance into consideration and assuming $\beta = \mu$ for the shear viscosity μ, it is notable that the Navier-St.Venant motion equation (2.25) for an incompressible model fluid ($\nabla \cdot v \equiv 0$) contains the Navier-Stokes motion equation as a 'pathological' limiting case (see STRAUB 1988, pp.141-144).

Seen from the standpoint of the AT, a notable conclusion can be made: for an incompressible fluid with constant shear viscosity μ, the Navier-Stokes equation of motion, together with the **incompressibility condition**

(2.30) $\nabla \cdot v \equiv 0$

and the compatible initial and boundary constraints, represents a complete set of equations. The solution – flow velocity v and pressure p dependent on location r and time t – is parametrically dependent only on the kinematic viscosity $\nu = \mu/\rho = $ constant (and, additionally, on the parameters fixed by the initial and boundary constraints) ; to a certain extent individualizes the incompressible fluid with a number for comparison.

In reality things look completely different: for all compressible fluids the new quantity ß (a physically corresponding function of μ) – equation (2.26) – is normally a function of state, provided the proportionality coefficient t_τ is either a constant or a known function of the density ρ, the mass fraction ω_k and the temperature T. In the latter case this characteristic time t_τ is a property of state like the pressure p, the compressibility factor Z and the absolute specific entropy s of the polynary single-phase system.

Even simple examples, however, lead one to expect that this coefficient ß is more likely to be dependent on the class of given flows (for example, a shear flow under certain constraints such as constant shear and isothermal fluid) than on the material properties (see STRAUB 1988, section 5.2). It is evident that this difficulty raises questions about conventional viscosity concepts. This is not the place, however, to present an answer to this problem, particularly since other authors have already made some pertinent attempts (see, for example, NETTLETON 1987). If necessary, there are no theoretical objections against accepting t_τ as an empirical coefficient for the present.

In the work just mentioned, NETTLETON (1987, p.276) carried out an estimation which indicates that one must expect extremely short times of an order of magnitude of $0(10^{-13}$ sec) for t_τ. A. Laesecke[50] examined the conventional concept of viscosity. Using Maxwell's definition $t_M :=$ η/p with the measured viscosity data η, he determined relaxation times t_M which partly confirmed this order of magnitude.

Generally speaking, one arrives at an interesting conclusion: **the Navier-Stokes equation of motion can be substantiated only for incompressible model fluids; for real compressible materials, one can only define it!**[51]

It is understandable that the Navier-St.Venant equation of motion at present cannot be used for actual projects. Currently available numerical procedures and computers are still too inefficient to handle the work. Therefore it is appropriate to search for a first non-trivial approximation of equation (2.25). The answer naturally depends on which approximations one can agree upon.

The **gas kinetics fundamentals of non-uniform and thermally-perfect gases** lead one to the conclusion that the most important prerequisite for a decisive simplification of the problem, the inequation

$$(2.31) \qquad\qquad \nabla T/T \ll 1 \quad,$$

may be valid for the profiles of local temperatures. In equation (2.31) ∇T relates to the absolute value of the local temperature gradients. If one ignores extreme cases, the condition (2.31) is still fulfilled within the linear dimensions of the order of magnitude $(V_m/N_A)^{1/3} \approx$ 33 Å [52]. This means that in most practical cases there are no relevant limitations.

In these cases, both the Navier-St.Venant equation of motion (2.25) and the energy field equation can be simplified further in the form

$$
\begin{aligned}
(2.32) &\quad c_V \rho DT = - p\nabla\cdot\mathbf{v} - \nabla\cdot\dot{\mathbf{q}}_F + \tfrac{1}{2}\mu_M(\nabla\cdot\mathbf{v})^2 \\
(2.33) &\quad \rho D\mathbf{v} = \rho f - \nabla p_V \\
(2.34) &\quad p_V := p - \tfrac{1}{2}\mu_M \nabla\cdot\mathbf{v} \\
(2.35) &\quad \mu_M := t_T p \\
(2.36) &\quad p = RT\rho
\end{aligned}
$$

The same can be done with the field equations of the mass fractions, not shown here.

Equation (2.33) agrees **formally** with the Euler equation of motion. Yet the 'pressure' p_V by no means corresponds with the thermodynamic pressure p according to equation (2.36). Equation (2.33) describes dissipative processes and can, like the Navier-Stokes equation of motion, fulfill no-slip boundary conditions. Analogies due to numerical viscosity effects obviously are the reason why the Euler equations of motion have recently been so successfully applied in industry for simulating real flows.

An **equation of motion of the Euler type** (and not the Navier-Stokes equation of motion!) is thus a first step toward describing real flow fields of compressible fluids. All field properties such as velocity **v**, specific field force **f**, mass density ρ, pressure p and temperature T are clearly defined by the Gibbs Fundamental Equation and related to one another.

If one accepts for $\dot{\mathbf{q}}_F$ the known proportionality to $\mathbf{\nabla}T$ according to the First Law of Fourier

$$(2.37) \qquad \dot{\mathbf{q}}_F := - \lambda\ \mathbf{\nabla}T = - \mathbf{\nabla}\cdot[c_v\ \mu_M\ f\ \mathbf{\nabla}T] \quad ,$$

that is equivalent to assume steady transfer processes (see STRAUB et al. 1987) in connection with the Fourier heat flow vector $\dot{\mathbf{q}}_F$. The coupling of heat conductivity λ with the **Maxwell viscosity** (analog to the known gas kinetic relationship $\lambda = f\ c_v\eta$ between λ and the dynamic viscosity η of non-uniform gases[53]) allows further simplification of equations (32) and (33), assuming continued constant material data. If one adds the well-known relations

$$(2.38) \qquad c_v/R = (\kappa - 1)^{-1} \quad ; \quad \Omega := p/\rho = RT$$

for perfect gases and ignores the quadratic $\mathbf{\nabla}\cdot\mathbf{v}$ term in equation (2.32) as well as the field force term in equation (2.33), then these two equations are immediately transformed – taking equation (2.31) into account (see STRAUB 1988, p.148) – into the forms

$$(2.39) \qquad D\Omega = - (\kappa - 1)[\Omega\ \mathbf{\nabla}\cdot\mathbf{v} - \tfrac{1}{2}\nu_M\ \mathbf{\nabla}\cdot\mathbf{\nabla}\ \Omega] \quad \kappa := c_p/c_v$$

$$(2.40) \qquad D\mathbf{v} = - \mathbf{\nabla}\ \Omega + \tfrac{1}{2}\ \nu_M\ \mathbf{\nabla}(\mathbf{\nabla}\cdot\mathbf{v}) \quad ; \quad \nu_M := \mu_M/\rho \quad .$$

They are of particular interest because they contain only two dependent variables – the flow velocity **v** and the 'scalar potential' Ω, which with constant material data can also be easily connected with the specific enthalpy of a perfect gas – as well as two parameters κ and ν_M. To solve them, one needs suitable initial and boundary conditions, but not the continuity equation: the latter can subsequently be used, when fields $\Omega(\mathbf{r}, t)$ and $\mathbf{v}(\mathbf{r}, t)$ are established, to calculate the density field $\rho(\mathbf{r}, t)$.

Equations (2.39) and (2.40) were first given by U. Nehring[54], who discussed their remarkable mathematical structures. Using them, he obtained valuable qualitative solutions for many characteristic examples through numerical integrations (see NEHRING 1984).

To summarize the presented concepts and results of the AT, it is best to relate them to the Navier-Stokes theory:

The AT is an inherently non-contradictory field theory of compressible fluids. It is a general set of procedural rules also suitable for complex flows of multicomponent single-phase fluid mixtures as special cases of a body-field system. The AT offers an impressive number of new physical insights which occasionally deviate significantly from traditional continuum theories.

The Navier-Stokes equation of motion is not an adequate relationship for calculating the velocity field of compressible fluids. The Navier-Stokes equations, as a collective definition of a set of field equations for the description of high enthalpy flows, for example, are heuristic extrapolations. They are based on a physical gas kinetic model justified at most by conceptual artistry. Their description on the level of the non-linear field equations is based on the hypothetical and paradox-laden assumption that compressible fluids are generally 'integrable systems' in the sense of Newtonian mass point mechanics.

> "We have stressed, with more than usual emphasis,
> the physics underlying the initial choice of variables
> of thermodynamic theory. That choice is a fateful one."
> —H. Callen—

3. The Munich Method

3.1 Preliminary Remarks

The existing qualitative results of the AT (in the form of field representations of the numerically integrated Nehring equations) are so promising that they deserve careful analysis, particularly with regard to
the increasingly important spatially and temporally complex high enthalpy flows. The fundamental elements of the AT serve as the basis for
the set of premises used for the **Munich Method (MM)**. The method's aim
is the assertible representation of a thermodynamic ideal comparative
process for relaxing flows (ICP).

The analysis is broken up into several stages, corresponding to the
discussion of the sequence of states introduced in Chapter 2 of Part I.

The MM is the terminus technicus for the mathematical representation of
a thermodynamically-based calculation procedure which can only be used
iteratively with respect to concrete problems. Its universality is
founded on the possibility of establishing a sequence of states (with
characteristic thermodynamic dependencies) which has relevance to the
design and construction of engines. All properties of state for every
point in the sequence of states thus have to be determined by a minimum
of information. In doing so, it is interesting to note that knowledge
of the geometric structure of the rocket engine (RE) can be reduced to
only a few details: detailed knowledge of the nozzle contour, in particular, is not necessary for the ICP.

3.2 Changes of State in the Combustion Chamber

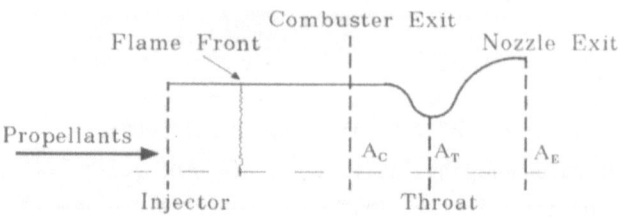

Figure 3.1: Configuration of a Combustion Chamber and Laval-Nozzle

In the above configuration, the thermofluiddynamic system – the ideal multicomponent single-phase fluid mixture resulting as a reaction product from the mixture of oxidizer O and fuel F – passes through the steady states $\underline{F}$, $\underline{C}$, $\underline{I}$ and $\underline{E}$. The realization of the process is isentropic.

One of the characteristic aspects of the MM is the calculation of the 'pressure drop' along the combustion chamber ($\underline{F} \rightarrow \underline{C}$) using the Alternative Theory. The constancy of the combustion chamber cross section area A_C is the determining factor. Accordingly, the transition from state $\underline{F}$ to $\underline{C}$ is limited by the constraint

$$(3.1) \qquad (\rho v)_F = (\rho v)_C \quad \text{resp.} \quad d(\rho v) = 0 \quad .$$

For steady, spatially one-dimensional processes, one can considerably simplify the **First Law for open systems** – for pure flows, without external work –

$$(3.2) \qquad \rho \, De_*^{[\rho]} + \mathbf{\nabla} \cdot \mathbf{j}_{e*}^{[\rho]} = 0 \quad ,$$

(see STRAUB 1988, p.70), if one agrees upon the required isentropy in 'open' flow cross section areas per

$$(3.3) \qquad \mathbf{\nabla} \cdot \mathbf{j}_{e*}^{[\rho]} \equiv 0 \quad .$$

In this expression, $\mathbf{j}_{e*}^{[\rho]}$ stands for the total **energy flow density** of the Legendre-transformed specific (total) energy

$$(3.4) \qquad e_*^{[\rho]} := e + \frac{p_*}{\rho}$$

for moving systems (* subscript). The specific total energy e of the system with all relevant properties of state is already known from equation (2.13).

Using an important relationship from the AT (STRAUB 1988, equation 91.2), one can now transform the Legendre-transformed specific energy $e_*^{[\rho]}$ to the corresponding energy $e^{[\rho]} := e + p/\rho$ of the multicomponent single-phase fluid mixture in a hypothetical state of rest. This transformed specific energy $e^{[\rho]}$ then can be divided up into various types of energy with

$$(3.5) \qquad\qquad e^{[\rho]} = h + e_{kin} + e_{pot} + e_{\phi} \qquad .$$

One arrives at the specific enthalpy h, the specific kinetic or potential energy e_{kin} and e_{pot}, and the specific dissipation energy e_{ϕ}. The latter is coupled with the dissipation velocity ϕper $e_{\phi} := -\frac{1}{2}\phi^2$ through equation (2.21), which itself constitutes the marked difference between flow velocity **v** and the specific impulse **i** of a real compressible fluid. This difference disappears only for dissipation-free flows (equation 2.24), so that in this case $e_{\phi} \equiv 0$ also is valid. If one eliminates the influence of potential energies, equation

$$(3.6) \qquad\qquad \rho\mathbf{v}\cdot\mathbf{\nabla}(h + \tfrac{1}{2}\mathbf{v}^2) = 0$$

is ultimately derived for steady flows using equation (3.2) and observing equation (3.3). Since in the one-dimensional case the mass flow density ρv is a constant, according to equation (3.1), the known relationship (compare equation 1.36)

$$(3.7.1) \qquad\qquad h_F + \tfrac{1}{2}\, v_F^2 = h_C + \tfrac{1}{2}\, v_C^2 \qquad ;$$

follows: it is valid only for frictionless changes of state (see ZIEREP 1976, p.30) and is equivalent to the condition

$$(3.7.2) \qquad\qquad de_*^{[\rho]} \equiv 0$$

for the Legendre-transformed specific (total) energy in a moving system. This condition guarantees that the 'environment' has no influence on the changes of state in a system; cooling processes, for example, thus are excluded (see Section 3.7).

The changes in the system itself are mathematically described by its Gibbs Fundamental Equation (GFE); it is obtained from equation (2.5) after being divided by the constant baryon mass and the subsequent Legendre transformation of the density variable to

$$(3.8) \qquad \rho \; de_*^{[\rho]} = \rho \mathbf{v} \cdot d\mathbf{i} + dp_* + T_* \rho \; ds + V^{-1} \sum_{k=1}^{K} \mu_{k*}^m \; dn_k \quad .$$

Contrary to equation (2.5), this fundamental relationship ignores the field force F and, in contrast to the Gibbs-Euler Function (2.13), introduces both the mole number $n_k := N_k/N_L$ as a variable and the molar chemical potential μ_{k*}^m (compare equation (1.14) which is exclusively valid for a mixture in a hypothetical state of rest). For the following considerations, it is useful to transform the sum term of equation (3.8) into a form appropriate for describing simultaneously-occurring chemical reactions. For this purpose two characteristic properties are introduced:

$$(3.9) \qquad dn_k := \sum_{r=1}^{\bar{R}} \nu_{kr} \; d\xi_r \qquad \text{-reaction coordinate } \xi_r -$$

$$(3.10) \qquad A_r := - \sum_{k=1}^{K} \nu_{kr} \; (\mu_{k*}^m/V) \qquad \text{-(chemical) affinity } A_r - \; .$$

The stoichiometric coefficients, previously defined by the reaction equation (1.7.1), are given with ν_{kr} . The index r refers to all chemical reactions taking part in the process – $r = 1(1)\bar{R}$ – and numbers their reaction equations.

With definitions (3.9) and (3.10), equation (3.8) is transformed to the expression

$$(3.11) \qquad \rho \; de_*^{[\rho]} = \rho \mathbf{v} \cdot d\mathbf{i} + dp_* + T_* \rho ds - \sum_{r}^{\bar{R}} A_r d\xi_r \quad ;$$

it is the reference point for one of the results essential to the MM: when equation (3.11) is solved for the energy form $T_* ds$ and equation (3.1) is taken into account, one obtains the form

$$(3.12) \qquad T_* \rho ds = \rho \; de_*^{[\rho]} - d[\rho \mathbf{v} \mathbf{i} + p_*] + \sum_{r}^{\bar{R}} A_r d\xi_r$$

which is the specialized GFE valid for the changes of state along a constant cross section area flow tube. It expresses the specific entropy s of the moving system as a function of the variables $e_*^{[\rho]}$, ξ_r, and $[\rho \mathbf{v} \mathbf{i} + p_*]$, too.

Such a relationship corresponds exactly, however, to the Y-function of state introduced in Section 1 of Part II. This function is dependent on the canonical variables χ_α^{can} ($\alpha = 1(1)A$) used for a generalized non-stoichiometric formulation of the equilibrium problem and originally involved in equation (1.38).

This equilibrium condition itself is expressed

$$(3.13) \qquad \left[\frac{\partial s}{\partial \xi_r}\right]_{e_*^{[\rho]}, \ \xi_{\neq r}, \ [\rho v i + p_*]} \overset{max.!}{=} 0 \ ; \qquad r = 1(1)\bar{R}$$

and is, among the constraints (marked as index), equivalent to the requirement for the affinity

$$(3.14) \qquad \hat{A}_r(s, \ e_*^{[\rho]}, \ \xi_r, \ [\rho v i + p_*]) \equiv 0 \quad ; \qquad r = 1(1)\bar{R} \quad .$$

Assuming a **reversible** adiabatic change of state, the specific impulse i equals the local flow velocity **v**, and pressure p_* equals pressure p of the thermal equation of state.

In a steady one-dimensional flow, the constraint $[\rho v i + p_*]$ = constant then simply corresponds with equation

$$(3.15) \qquad p_F + (\rho v^2)_F = p_C + (\rho v^2)_C \quad ,$$

which is identical with the flow tube theory's motion equation for constant flow tube cross section areas (see equation 1.35).

The constraint $e_*^{[\rho]}$ = constant for dissipation-free, steady one-dimensional flows is already expressed in the First Law for open systems as Eq.(3.7.2). Equation (3.13) therefore establishes the chemical equilibrium not subject to the usual constraints such as T = constant, p = constant (i.e. $\hat{G}(T,p,\xi_r)$: minimum) or h = constant, p = constant (i.e. $\hat{s}(h, p, \xi_r)$: maximum), but rather to constraints for total properties relevant to the flow tube theory (see PENNER 1957, pp.120 f).

The result allows an interesting physical interpretation: as long as chemical equilibrium can be reached at least in limiting cases, the values of reaction coordinate ξ_k^{equ} or of mole number n_k^{equ} of the k^{-th} component are directly influenced by the flow. They differ from those values found in the system's otherwise comparable state of rest (equation 2.8). It is evident that this result cannot be obtained without incorporating the energy form of the motion $\mathbf{v} \cdot d\mathbf{P}$ in a system's GFE!

Attention should also be called to contributions made to an old controversy over the problem of chemical equilibria in flows. The solution obtained in equation (3.14) indicates that such an equilibrium can be **proven** only in the special case treated here. Equilibria certainly cannot be represented in real flows with the usual equilibrium constants (see JISCHA 1973, PETERS 1975, and VINCENTI & KRUGER 1967).

The required function of state $Y = \hat{Y}[e^{[\rho]}, (\rho v^2 + p), n_k]$ for the 'reversible adiabatic flow in a tube with a constant cross section area' thus exists for the formulation of the sought Lagrange function $\mathscr{L}_s$. Condition (3.13) in the equilibrium relationship

$$(3.16) \qquad \left[\frac{\partial s}{\partial n_k} \right]_{h^{tot}, \, p^{tot}, \, n_{\neq k}} \overset{max.!}{=} 0 \; ; \quad k = 1(1)K$$

can be rewritten directly over equation (3.8). As usual, the 'total' values of the properties of state h and p can be defined by the sums occurring in equations (3.7.1) and (3.15). The prerequisites for using equation sets (1.38) or (1.39) are given by equation (3.16).

If one confines the study to the case of a rocket engine in which energy condition $\frac{1}{2}v_F^2 \ll \frac{1}{2}v_C^2$ is considered fulfilled, the general scheme (1.39) is right. It corresponds strikingly with Prozan's (1.41) postulated scheme for calculating the mole numbers with corresponding temperature and pressure data in chemical equilibria – as long as one does not take the Prozan function ϕ^T into consideration and sets the Lagrange multiplier λ^T at zero.

Proper arrangement of this set (1.41) of non-linear algebraic equations for numerical treatment requires complex preparation aimed at incorporating the variety of relevant physical information into the mathematical modeling. One begins with the Lagrange function for scheme (1.41), naturally for $\lambda^T \equiv 0$:

$$(3.17) \qquad \mathscr{L}_s := - s + \sum_e \lambda_e (b_e^{\varnothing} - \sum_k a_{ek}\, n_k) +$$

$$+ \lambda^E (h + \frac{b}{2}\, v^2 - h_F) + \lambda^M (p + bv - p_F) \quad .$$

Equations (1.30) and (1.31), (or alternatively Eq.(1.22)) are valid respectively for the mixture's specific entropy s and enthalpy h.

The known mole number $b_e^{\varnothing}$ of the e^{-th} element in bounded or unbounded states for the chosen reference state is defined with Eq.(1.10.1). The specific volume v as well as the squared mass flow density b (treated as a parameter) have already been introduced with equation (1.37).

At first the specific enthalpy h_F is established by equation (1.28) for the given parameters p_F and O/F (in the case of the LH-LOX combustion: O/F = $\dot{m}_{LOX}/\dot{m}_{LH}$). Then, the fluid mixture's specific entropy s_F may be determined 'in principle' by equation (2.b) of Appendix 2. In practice, however, the initial values T_{ad} and $x_{k,ad}^{equ}$ calculated with the AFC method yield the specific entropy s_{ad} as a first approximation for s_F. In addition, a value for b := $(\dot{m}/A_c)^2$ and thus for the steady mass flow rate $\dot{m}$ is estimated.

With these informations scheme (1.41) can be computed. The required conditions for all elements e = 1(1)2 and components k = 1(1)6 are (with respect to the LH-LOX-combustion) explicitly formulated by differentiations of the Lagrange function $\mathscr{L}_s$ according to the respective variables:

$$\partial\mathscr{L}_s/\partial n_k = 0 \quad ; \quad \partial\mathscr{L}_s/\partial T = 0 \quad ; \quad \partial\mathscr{L}_s/\partial p = 0$$

$$\partial\mathscr{L}_s/\partial\lambda_e = 0 \quad ; \quad \partial\mathscr{L}_s/\partial\lambda^E = 0 \quad ; \quad \partial\mathscr{L}_s/\partial\lambda^M = 0 \quad .$$

These calculations are elementary and need not be presented here. Given is the set of equations, whose solution considering equation (1.42) allows the calculation of all properties of state in state $\underline{C}$ [55]

(3.18)

$$
(1)-(6) \quad g_k + \psi_k^{-1} T[\lambda_0\, a_{0k} + \lambda_H\, a_{Hk}] + \phi_k^E = 0 \quad ; \quad k = 1(1)6
$$

$$
\phi_k^E := \overline{\lambda^E}\left[h_k + (1 - x_k)T\,\frac{dh}{dT}\right]
$$

$$
\overline{\lambda^E} := \lambda^E T - 1 := \theta
$$

$$
(7) \quad \sum_k \left[a_{0k} - a_{Hk}\,\frac{\psi_H}{\psi_0}\,(O/F)\right] x_k = 0 \quad ; \quad (8) \quad \sum_k x_k = 1
$$

$$
(9) \quad h - h_F + \tfrac{1}{2}b(RT/p)^2 = 0 \quad ; \quad R := \mathbf{R}/\psi
$$

$$
(10) \quad p - p_F + b(RT/p) = 0 \quad ; \quad \psi := \sum_k \psi_k x_k
$$

$$
(11) \quad s - s_F = 0
$$

.

The equation set (3.18) consists of 11 equations for the following 11 properties of state in state $\underline{C}$: pressure p and temperature T, six mole fractions x_k, $k = 1(1)6$, and the three Lagrange multipliers λ_0, λ_H and θ (= LAMBDA-E-BAR, see the protocols in Section 4.2). The specific free enthalpy of the k^{-th} component is denoted with g_k.

It should be explicitly noted here that the solution of equation set (3.18) requires knowledge of the properties h, p and s in the **fluid** state $\underline{F}$. For the peculiar circumstances in the LH-LOX combustion, approximations must be used for determining the specific fluid enthalpy h_F and entropy s_F. Such approximations are given in Appendix 2, together with some explanations at the end of Section 3.6.

It is apparent that the Lagrange multiplier λ^M does not appear in equation (3.18). As is easily shown, there is a direct relationship

$$
(3.19) \qquad \lambda^M = -\frac{1}{p}\left[R + \overline{\lambda^E}\,\frac{dh}{dT}\right] \quad ;
$$

between λ^M and θ for a non-trivial solution (i.e. for $\lambda^E \neq 1/T$) of the equation set.

The first relationship of the equation set (3.18) is of particular interest. It clearly illustrates one of the most striking differences between the NASA-Lewis Code and the MM. In the case of $\lambda^E = 1/T$, (i.e. the trivial case $\theta \equiv 0$), the auxilliary function ϕ_k^E disappears for all components and the basic equation of the equilibrium calculation results: the minimization of the free enthalpy $\hat{G}(T,p,n_k)$ – equation (1.4) – at constant values for the constraints of pressure and temperature.

For the combustion chamber flows investigated in this study, negative θ-values of the order $O(10^{-2})$ appear. If one expands the formula $\lambda^E :=$ $(T + \Delta T)^{-1} \approx T^{-1}(1 - \Delta T/T)$, the simple relationship

$$(3.20) \qquad \theta := \overline{\lambda^E} \approx - \Delta T/T$$

is obtained for the modified parameter θ. Taking a definite gas temperature T in the combustion chamber as a basis, one can estimate the temperature difference ΔT as a reference for the deviation from the trivial case. With respect to the standard equilibrium it may be interpreted as the consequence of the altered constraints resulting from the steady flow rate of a mixture. Given the temperature $T \approx 3000$ K, one obtains, for example, values of approximately $\Delta T \approx 30$ K. Such values naturally are noticeable in an exponential dependency of the temperature influence on chemical reactions.

With the resolution of the set of equations (3.18), all obtainable informations in state $\underline{C}$ are known. Mass density ρ as well as mean molar mass ψ , for example, are derived from the physical system's thermal equation of state, equation (1.15) and (1.23.2). Since the mass flow rate $\dot{m}$ is treated as a parameter for a known cross section area A_C, the gas velocity v_C as well as the Mach number $M_C = v_C/a_C$ are fixed. In order to calculate the latter number, the **speed of sound** a_C of the chemically-reacting gas mixture must be known. Its determination will be treated in the following section.

In view of the strongly non-linear equations, the set's own resolution needs as a rule sophisticated solver procedures like Gordon & McBride's efficient equilibrium calculation code. This is especially true for chemical equilibrium determinations with oxidizer-fuel mass flow ratios for fuel-poor combustion.

3.3 Speed of Sound in Chemically-reacting Gas Mixtures

Under the given circumstances, the properties of state in $\underline{C}$ can be fixed for the combustion gas. Direct proof of the thermodynamically consistent calculation of the equilibrium composition is based on the AT.
This is a decisive prerequisite for the **quantification** of the sequence
of states, especially in state $\underline{I}$.[56] This state, which is 'dynamically`
defined for the ICP (i.e. through the Mach number condition M = 1),
does not coincide simultaneously in a real nozzle reactive flow with
the state in the nozzle's geometric throat (see PIRUMOV & ROSLYAKOV
1986, pp.214 f). In the case of the ICP's chemically-reacting gas mixture along a **flow tube with variable cross section area**, the opposite
is true:

the Mach number condition

(3.21) $$M_{\underline{I}} \equiv 1$$

holds in the nozzle throat.

This is primarily due to the fact that the equation of motion for a
compressible polynary single-phase system with or without steady conversion processes corresponds with the Euler equation of motion if the
processes occur reversibly and adiabatically. Together with the condition for constant steady mass flow rate $\dot{m} := A\rho v$ which constitutes the
flow tube, differential relationships

(3.22.1) $$d(A\rho v) = 0$$

(3.23) $$\rho v dv + dp = 0$$

are obtained for isentropic processes, given the restriction that mo-
mentum-, energy-, or mass-transports be absent.
In other words: the ICP is simply an ordered succession of equilibrium
states, whereas a real process is a temporal succession of non-equilibrium thermodynamic states (see CALLEN 1966, p.60).

Nozzle gas flows are characterized by gasdynamic parameters with large
gradients. There are limiting values for relaxation times useful for
comparison with reactive flows close to equilibrium or near frozen

states (see Figure 4.1 in Part I). Intermediate states are strongly influenced by the locally dominating process realization. These facts are also significant for the speed of sound term to be discussed below.

Equation (3.22.1) can be expressed in the form

(3.22.2) $$\frac{dA}{A} + \frac{d\rho}{\rho} + \frac{dv}{v} = 0$$

and then dA/A resolved at

(3.22.3) $$\frac{dA}{A} = - \frac{dv}{v} \left[1 + \frac{v}{\rho} \frac{d\rho}{dv}\right]_s \quad .$$

This expression allows an important connection with equation (3.23)

(3.24) $$\frac{dA}{A} = - \frac{dv}{v} \left[1 - v^2 [\frac{\partial p}{\partial \rho}]_s^{-1}\right]$$

and leads to some relevant conclusions:

it is evident that if state $\underline{I}$ can be assigned not only a geometric property such as area $A_{\underline{I}}$, but the condition

(3.25) $\qquad$ dA/A = 0 $\quad$ (narrowest flow tube cross section)

as well (which is equivalent to a formal characterization of the nozzle throat), then according to equation (3.24), it is

(3.26) $$v_{\underline{I}}^2 = [\frac{\partial p}{\partial \rho}]_s := a^2 \quad ;$$

this characteristic velocity $v_{\underline{I}}$ is called the **speed of sound**. It results as a property of state primarily from the dissipation-free equation of motion. This steady, one-dimensional Euler equation can be derived from the Boltzmann collision equation for the special case of Maxwell's concept of **kinetic equilibrium**; it requires, above all, reversibility, i.e. disappearing entropy production density σ but not disappearing gradients of all field variables!

This circumstance is of particular importance because the differential expression $(\partial p / \partial \rho)_s := a^2$ usually is either simply defined as the speed of sound or 'derived' with more or less dubious thermostatic arguments (see LIEPMANN & ROSHKO 1957, pp.69 f). But such arguments cannot explain why a differential quotient, because it merely happens to correspond to the square of a value with a velocity unit, should be identical with the propagation velocity of sound without dissipative changes. Nor is it easy to see that such a 'derivation' can incorporate chemical processes, or how the individual physical facts in a reactive flow (such as stationariness as a constraint) can even be taken into consideration. Since equation (3.26) should be, strictly speaking, replaced by $a^2 := (\partial p/\partial \rho)_{\sigma=0}$, even in an idealized cooling process the velocity in $\underline{I}$ differs from $v_{\underline{I}}$ in an isentropic flow (see Section 3.7).

The interpretation presented here also makes it immediately obvious that equation (3.26) by no means applies to **real nozzle reactive flows**: i.e. $v_{\underline{I}} \neq a$ is valid as long as one considers the quantity a to be the differential expression as a definition and not the speed of sound.[57]

Whereas the calculation of a is simple for a perfect gas and leads to the known result $a = (\kappa RT)^{\frac{1}{2}}$, the procedure is rather complicated for an ideal polynary single-phase system. This calculation is presented below in all its relevant parts.

For simplicity's sake, the differential quotient $a^2 := (\partial p/\partial \rho)_s$ assumes an equation of state $p = \hat{p}(\rho, s, n_k)$ with the variables ρ, s and n_k , k = 1(1)K for a reactive fluid mixture. Since a is supposed to be dependent on variables T and p for the MM, however, a complex formulation is unavoidable.

As will be demonstrated shortly, it is practical to use the equivalent relationship

$$(3.27) \qquad\qquad a^2 = \left[\frac{\partial p}{\partial T}\right]_s / \left[\frac{\partial \rho}{\partial T}\right]_s$$

as an exact substitute for equation (3.26).

To calculate the differential quotient $(\partial \rho/\partial T)_s$, the energy form of motion in the GFE (3.8) can be easily separated for dissipation-free flow processes with the identity $\phi \equiv 0$ for the dissipation velocity. The result thus corresponds with the thermodynamic relations of the hypothetical state of rest. With $e_{pot} \equiv 0$ and $e_{\phi} \equiv 0$, and using equation (3.5), one can choose the specific enthalpy h as dependent property of state rather than the Legendre-transformed specific energy $e^{[\rho]}$. In addition,

only under these circumstances is the thermal equation of state a valid
tie between the three thermal variables of state T, p and ρ as well as
the mean molar mass ψ on the one hand, and the total mole number n of
the ideal mixture on the other. Expressed as a differential, this relationship for a steady mass flow is

(3.28) $$dp = \psi^{-1} RT\rho \left[\frac{dT}{T} + \frac{d\rho}{\rho} + \frac{dn}{n} \right]$$

with R as universal gas constant.

Using equation (3.12), the expression

(3.29) $$T\rho ds = \rho \, dh - dp + \sum_r A_r \, d\xi_r \quad ; \quad r = 1(1)\bar{R} \quad ,$$

holds for the hypothetical state of rest. This well-known standard
Gibbs identity with the required constraint ds = 0 is simplified when
the process realization occurs in chemical equilibrium. In this case,
the sum in equation (3.29) disappears and the reaction coordinates ξ_r ,
which are representative for the $\bar{R}$ reaction equations, can be expressed
as dependents only of T and p. This satisfies the intention of describing the speed of sound in these variables.

Equation (3.29) is vital for this problem and can be simplified to the
form

(3.30) $$dH = V \, dp \quad ; \quad V = n \, \psi \, \rho^{-1} \quad .$$

This relation serves for precise calculating the differential quotient
$(\partial \rho / \partial T)_s$. It is especially useful, since the enthalpy H can easily be
shown as a function of T and p, also in the state of equilibrium.
The enthalpy $\hat{H}(T, p, n_k)$ has the total differential

(3.31) $$dH = \left[\frac{\partial H}{\partial T} \right]_{p,n_k} dT + \left[\frac{\partial H}{\partial p} \right]_{T,n_k} dp + \sum_k h_k^m \, dn_k \quad ,$$

in which the prefactors concerning an ideal gas mixture are defined by

(3.32) $$n \, c_{p,\xi}^m = \left[\frac{\partial H}{\partial T} \right]_{p,n_k} \quad ; \quad \left[\frac{\partial H}{\partial p} \right]_{T,n_k} \equiv 0 \quad ; \quad k = 1(1)K \quad ,$$

and the partial molar enthalpy h_k^m is dependent only on T.

Using equation (3.9), substitution of the variable n_k with reaction co-ordinates ξ_r allows the differential $d\xi_r$ in an equilibrium state to be expressed through changes of the variables T and p:

$$(3.33) \qquad \xi_r = \hat{\xi}_r^{equ}(T,\ p) \rightarrow \frac{1}{n}\ d\xi_r^{equ} = \xi_{r,T}^m\ dT + \xi_{r,p}^m\ dp \quad .$$

The prefactors $\xi_{r,T}^m$ and $\xi_{r,p}^m$, $r = 1(1)\bar{R}$, stand as abbreviations for the reaction coordinate's local temperature and pressure derivatives in an equilibrium state at T and p. HOLUB & VOŇKA (1976, Section 4.3.1) have given the relationship

$$(3.34) \qquad \boxed{\ \xi_{r,T}^m = \frac{1}{T}\ \frac{\sum\limits_k \nu_{kr}\ [h_k^m / RT]}{\sum\limits_k \left[\nu_{kr}^2\ [x_k^{equ}]^{-1} - \nu_r^2\right]} \quad ; \quad r = 1(1)\bar{R}\ }$$

for $\xi_{r,T}^m$.
In addition, the expression

$$(3.35) \qquad \boxed{\ \xi_{r,p}^m = \frac{1}{p}\ \frac{\nu_r}{\sum\limits_k \left[\nu_{kr}^2\ [x_k^{equ}]^{-1} - \nu_r^2\right]} \quad ; \quad \nu_r := \sum\limits_k \nu_{kr}\ }$$

is valid for the pressure derivative.

The denominator obviously is always positive in both equations. There-fore, an increase or decrease of temperature or pressure dependency of every reaction coordinate is determined primarily by the sign and abso-lute value of the molar reaction enthalpy $\sum \nu_{kr} h_k^m$ or **reaction order** ν_r of the r-th reaction. Both prefactors are exclusively dependent on T and p, since the mole fraction x_k of the k-th component in chemical equilibrium is also influenced only by these variables.

Furthermore, using equations (3.31) to (3.33) while observing equation (3.28) and the total derivative of the mole number

$$(3.36) \qquad \frac{1}{n}\ dn = dT \sum_r \xi_{r,T}^m\ \nu_r + dp \sum_r \xi_{r,p}^m\ \nu_r \quad ,$$

equation (3.30) becomes, after some manipulations,

(3.37)
$$\left[\frac{\partial \rho}{\partial T}\right]_s = \frac{\rho}{T} \left[\left[\frac{c_{p,\xi}^m}{R} + T \sum_r \xi_{r,T}^m \sum_k \nu_{kr} \frac{h_k^m}{RT} \right] \frac{1 - p \sum_r \xi_{r,p}^m \nu_r}{1 - p \sum_r \xi_{r,p}^m \sum_k \nu_{kr} [h_k^m/RT]} - \right.$$

$$\left. - \left[1 + T \sum_r \xi_{r,T}^m \nu_r \right] \right]$$

for the differential quotient $(\partial \rho/\partial T)_s$.

In contrast, the differential quotient $(\partial p/\partial T)_s$ is derived directly from equations (3.30) to (3.33). For its representation, characteristic abbreviations are used for terms which appeared in equation (3.37):

(3.38.1) $$\Gamma_T := T \sum_r \xi_{r,T}^m \sum_k \nu_{kr} \frac{h_k^m}{RT} \quad ; \quad \Gamma_p := p \sum_r \xi_{r,p}^m \sum_k \nu_{kr} \frac{h_k^m}{RT}$$

(3.38.2) $$\Xi_T := 1 + T \sum_r \xi_{r,T}^m \nu_r \quad ; \quad \Xi_p := 1 - p \sum_r \xi_{r,p}^m \nu_r$$

This yields the derivative

(3.39) $$\left[\frac{\partial p}{\partial T}\right]_s = \frac{p}{T} \left[\frac{c_{p,\xi}^m}{R} + \Gamma_T \right] (1 - \Gamma_p)^{-1} \quad ,$$

an expression which is considerably simpler than equation (3.37). This equation is of special significance as **"nozzle differential equation"** for the MM.

To provide a particularly transparent explanation of the speed of sound for a multicomponent single-phase fluid mixture in chemical equilibrium, it appears practical first to reach an agreement in accordance with the classical result:

(3.40) $$a := (\gamma \, R \, T)^{\frac{1}{2}} \quad ; \quad R := R/\psi \quad .$$

Comparing this with definition (3.26) and using equations (3.27),

(3.37) and (3.39), one finally obtains

$$(3.41) \quad \boxed{\gamma = \frac{\Gamma_T + c^m_{p,\xi}/R}{[\Gamma_T + c^m_{p,\xi}/R]\, \Xi_p - [1 - \Gamma_p]\, \Xi_T}}$$

for the **modified isentropic exponent** (see BARRèRE et al. 1960, pp.175 - 178). The explicitly formulated derivatives (3.34) and (3.35) primarily guarantee a simple and reliable calculation of this exponent.

In the critical cases

(i)   **no** chemical reactions: $r = 1(1)\bar{R}$

$$\xi_r \equiv 0 \;\longrightarrow\; \begin{cases} \xi^m_{r,T} \equiv 0 \\[4pt] \xi^m_{r,p} \equiv 0 \end{cases} \;\longrightarrow\; \gamma = \kappa := c^m_p/(c^m_p - R)$$

(ii)   **frozen** chemical reactions: $r = 1(1)\bar{R}$

$$\xi_r = \text{constant} \;\longrightarrow\; \begin{cases} \xi^m_{r,T} \equiv 0 \\[4pt] \xi^m_{r,p} \equiv 0 \end{cases} \;\longrightarrow\; \gamma = \gamma_f := c^m_{pf}/(c^m_{pf} - R)$$

this property of state γ actually is solved correctly; the index f for 'frozen reaction' marks a process in which changes of state occur at unchanged (compared to the initial values) values of the gas mixture's mole fractions. Such processes in rocket engines, especially in diverging parts of the Laval nozzle, often influence the main performance data (see K.R.C. Bray in WEGENER (Ed.) 1970, pp.147 f).

3.4 Stoichiometric Matrix of the LH-LOX Equilibrium Combustion

The result (3.41) illustrates the substantial complications which immediately arise when chemical reactions occur in a gas mixture. These complications contrast with both classical flow tube theory and simplified reactive gas dynamics! It is still possible, however, to derive a closed algebraic formula for the speed of sound a. In doing so, it is notable that a separate summation with the index r is required for all

auxiliary functions (3.38). In addition, the stoichiometric coefficients ν_{kr} appear. Identifying them requires that the gas components and
reaction mechanisms of the combustion process be correlated for the entire sequence of states. This is a characteristic situation for equilibrium calculations using the stoichiometric formulation of all equilibrium conditions (see, for example, SMITH & MISSEN 1982, p.45). Important pertinent details were previously covered in Section 1.2. In
this particular context, one may recall that in chemical equilibria,
the number of all chemical equations $\bar{R}$ to be taken into consideration
must equal the number R of all **linear independent chemical equations**.
In this case $\bar{R}$ is the rank of the stoichiometric matrix $\underset{\sim}{\nu}$ (compare
equation (1.7.2)). Together with the formula matrix $\underline{A}$, matrix $\underset{\sim}{\nu}$ forms
the matrix equation (1.9): it is decisive that $\underline{A}$ belongs to the system
definition. This fact essentially means that not only the total number
K of mixture components, but also the chemical nature of each component
is part of a given model, considering the chemical elements bounded.
This information is necessary for establishing the integers a_{ek} constituting the formula matrix $\underline{A}$ (see commentary on equation (1.9)).
It must be emphasized that the set of chemical equations corresponding
to the stoichiometric matrix $\underset{\sim}{\nu}$ – see the equation set (1.7.1) – can by
no means be clearly ascribed to a realistic model of the chemical mechanism involved:
"It is generated solely from the list of species presumed (or demonstrated) to be present, that is, from $\underline{A}$, and neither requires nor implies any knowledge of reactions presumed to be taking place, or of reaction mechanisms." (SMITH & MISSEN 1982, p.22).

The situation described above can be effectively demonstrated with the
example of the hydrogen combustion treated in this study. The LH-LOX
reaction mechanism important for reaction-kinetic fundamentals has been
investigated for over 20 years in context with the development of upper-stage rocket engines. Throughout this period combustion chamber
pressures have been conspicuously raised to ever higher levels although
knowledge of the pressure's influence on reaction-kinetic elementary
effects still must be considered rudimentary at best (see WILHELMI
1987). Quite recently, E.U. Franck was able to demonstrate experimentally the steady flame of a 1000 bar H_2-O_2-combustion and its phenomenological behaviour (see FRANCK 1987, pp.225 f).

The H_2-O_2-combustion is a typical example of a chain-branching reaction (see GARDINER 1972, pp.214 f). If one studies, for example, the early investigations conducted by R.F. Buswell, V.J. Sarli and their associates, which were the basis of the NASA performance calculations and design work for sub- and supersonic ramjet combustion (see FRANCISCUS & HEALY 1967, p.6), one can clearly see the increasingly detailed experimental and mathematical attention paid to the elementary reactions occurring in combustion systems. Such studies made careful comparative investigations of flame velocity, the maximum values of the mass fractions of H, O, OH, H_2O, energy conversions, etc. (see J. Warnatz in PETERS & WARNATZ (Eds) 1982, pp.49 f). Yet in spite of an unexpectedly good congruence of some results, serious doubts remain about the validity of the current bases of calculation (see also D.M. Smooke et al. in PETERS & WARNATZ (Eds) 1982, pp.122-123) and the numerical computations for the underlying chemism.

In H_2-flames, H_2 is normally oxidized to H_2O exclusively in the reaction

$$\text{(I)} \qquad OH + H_2 \rightarrow H + H_2O \quad .$$

The governing (reaction) rate constant k_r is usually quantified by the ansatz — see e.g. PENNER 1957, pp.216 f —

$$\text{(3.42)} \qquad k_r := AT^b \exp[-\ E/RT]$$

as function $\hat{k}_r(T)$. The constant A has the unit $(cm^3/mol)^{\nu-1}\ s^{-1}$ (with ν as the **overall** or **reaction order**) and the property E (as a kind of activation energy) has the unit kJ/mol; b is a pure number. The following table compares the A, b and E data from five current reaction schemes for the reaction equation given above:

		A	b	E
(1) BAULCH et al.	1972/76	$2.2 \cdot 10^{13}$	0	21.5
(2) PETERS & WARNITZ	1982	$1.2 \cdot 10^{9}$	1,3	15.2
(3) GARDINER	1983	$1.0 \cdot 10^{8}$	1,6	13.8
(4) PIRUMOV&ROSLYAKOV	1986	$1.1 \cdot 10^{14}$	0	8.6
(5) BASEWICH	1987	$2.4 \cdot 10^{13}$	0	21.8

Scheme (1) was proposed by J. Warnatz, and scheme (3) by G. Dixon-Lewis and associates. It is difficult to decide under which circumstances which of the five available numerical sets is the most reliable.

The extensive literature on the H_2-O_2-combustion mechanism cannot be commented here, particularly since competent summaries are available (see, for example, BAULCH et al. 1972/76, and DIXON-LEWIS & WILLIAMS 1977).

The concurrent situation between the known chain-branching reaction O_2+ H $\rightarrow$ O + OH and the recombination reaction O_2+ H + M $\rightarrow$ HO_2+ M is apparently significant for the maintenance of the reaction (I): it can cause a break in the chain reaction especially at high pressures and relatively low temperatures. At T < 1200 K, however, the branching reaction H_2O_2+ M = 2OH + M (see BASEWICH 1987, p.213) dominates (M stands for any given collision partner). WILHELMI (1987) points out the relative relevance of triple collision reactions at high combustion pressures.

If one wants to perform reliable flame calculations for the H_2-O_2-reaction mechanism at chemical non-equilibrium, especially at high temperatures, one must incorporate the **polynary** transport coefficients as well as the thermal surface **radiation**. As a relevant example HEITMEIR (1987) and GRAUE (1988) have investigated the heterogeneous C-N-O reaction mechanism (see WILLIAMS 1985, pp.48 f). Their extensive theoretical-numerical and experimental results have been recently published.

Large numbers of reaction equations characterize such reaction-kinetic calculations. At present, reaction mechanisms chosen for the H_2-O_2-combustion range in number from 16 reaction equations proposed by J. Warnatz (p.51), to 11 by D.L. Baulch et al. (p.52) and to 26 by G. Dixon-Lewis (p.184) (the page notations refer to the workshop report edited by PETERS & WARNATZ 1982). WILHELMI (1988, pp.20-21) recently compiled a scheme with 46 reaction equations!

Such reaction mechanisms are only of limited value for the equilibrium equations being considered here. The information gathered from each set of reaction equations is certainly valuable: one can determine which species are to be considered as mixture components during the process with changes of temperature and pressure. This decision about number

and type of species is elementary for all thermofluiddynamic calculations. In addition, **the reaction scheme for equilibrium calculations
has a more fundamentally different physical significance than with investigations in reaction kinetics.** This difference is scarcely treated
in technical publications although it is of particular practical relevance. The reason for this general indifference lies in the fact that
reaction kinetics is primarily based on physico-chemical methods while
equilibrium calculations are of thermodynamic nature. Only in the past
15 years have the latter been subjected to rigorous treatment with matrix algebra. This has previously been referred to in Section 1.2.

It would be far too complex here to justify each individual equilibrium
reaction scheme for the H_2-O_2-combustion. Fortunately R.A. Strehlow has
already completed such a study. Since his analysis (STREHLOW 1984, pp.
100 f) is the only one easily available to date, its important results
need to be noted and briefly commented.

In his highly recommendable text, R.A. Strehlow treats the reactive
system that is also of interest here:

(3.43) $\{H_2, O_2, H_2O, H, O, HO; \mathbf{H, O}\}$.

The system consists of the given **six species as mixture components;**
they are either identical with the two elements **H** and **O** or contain them
in a bounded form.

If one presumes, for instance, the following six reversible reaction
equations (marked with $\rightarrow \leftarrow$ as an arrow symbol) to be **representative** of
the trend to chemical equilibrium between the mixture components at
given thermal conditions – compare the scheme in WILLIAMS 1985, p.574 –

$$H_2 + O_2 \;\rightarrow\leftarrow\; 2\,OH$$

$$H_2 + \tfrac{1}{2}O_2 \;\rightarrow\leftarrow\; H_2O$$

$$H_2 \;\rightarrow\leftarrow\; 2\,H$$

(3.44) $$O_2 \;\rightarrow\leftarrow\; 2\,O$$

$$H + HO \;\rightarrow\leftarrow\; H_2 + O$$

$$O + H_2 \;\rightarrow\leftarrow\; OH + H \quad ,$$

then this scheme is inadequate for the **calculation** of the equilibrium
composition. Following the explanation in Section 1.2 and the initial

analysis of both the derivatives $\xi_{r,T}^m$ and $\xi_{r,p}^m$ for the $\tilde{R}$ independent reactions, it is clear that scheme (3.44) cannot be correct, at least mathematically. According to equation (1.7.3), with $K = 6$ and $E = 2$, there are only $\tilde{R} = 4$ independent reaction equations permissible.

R.A. Strehlow clearly demonstrates how the correct equilibrium reaction scheme, representing the independent formation of the species OH , H_2 , O_2 and H_2O from the elements H and O, can be established. This scheme

(3.45)

$$
\begin{array}{l}
OH \quad \rightarrow \leftarrow \quad H + O \\
H_2 \quad \rightarrow \leftarrow \quad 2\,H \\
O_2 \quad \rightarrow \leftarrow \quad 2\,O \\
H_2O \quad \rightarrow \leftarrow \quad 2\,H + O
\end{array}
$$

differs from scheme (3.44) not only in the number of the reaction equations, but also in the physico-chemical content of at least the first and last reaction equations. It yields the obligatory stoichiometric coefficients ν'_{kr}, $k = 1(1)K$; $r = 1(1)\tilde{R}$ for the equilibrium calculation. The mass-balance equations

(3.46.1) $b_O^\emptyset = n_O + n_{OH} + 2n_{O_2} + n_{H_2O}$

(3.46.2) $b_H^\emptyset = n_H + n_{OH} + 2n_{H_2} + 2n_{H_2O}$

are allocated to scheme (3.45): in these balances the set of numbers in front of each mole number n_k forms the formula matrix $\underset{\sim}{\mathbf{A}}$.

The most striking result of Strehlow's analysis, however, is his proof that the contents of the equilibrium reaction scheme are influenced by the actual initial composition of the reactive system and thus by the reference state mole number $n_k^\emptyset$. The element mole numbers $b_e^\emptyset$ naturally cohere with the $n_k^\emptyset$, yet the $n_k^\emptyset$ often are given invariants for the equilibrium calculations, and can be directly related to the $b_e^\emptyset$ at best.

First the given initial mole numbers $n_{OH}^\emptyset$ and $n_O^\emptyset$ for the reactive system defined by the set (3.43) have to be examined. For this example the mass balance equations

(3.47.1) $\qquad n_{OH}^{\emptyset} = n_{OH} + 2n_{H_2O} + n_H + 2n_{H_2}$

(3.47.2) $\qquad n_0^{\emptyset} = n_0 - n_{H_2O} - n_H - 2n_{H_2} + 2n_{O_2}$

hold.

To this set the equilibrium reaction scheme is assigned

$$O + H_2O \;\rightleftarrows\; 2\,OH$$

$$O + H \;\rightleftarrows\; OH$$

(3.48) $\qquad O_2 \;\rightleftarrows\; 2\,O$

$$2\,O + H_2 \;\rightleftarrows\; 2\,OH$$

which considerably differs from scheme (3.45), and thus also has a different stoichiometric matrix $\underline{\nu}$.

In the second example, the mole numbers of O_2, H_2 and H_2O are given as
initial data. The mass balance equations

(3.49.1) $\qquad n_{O_2}^{\emptyset} = n_{O_2} + \tfrac{1}{2}\, n_H + n_0 + \tfrac{1}{2}\, n_{OH}$

(3.49.2) $\qquad n_{H_2}^{\emptyset} = n_{O_2} + \tfrac{3}{2}\, n_H + n_0 + \tfrac{1}{2}\, n_{OH}$

(3.49.3) $\qquad n_{H_2O}^{\emptyset} = n_{H_2O} - n_H - n_0$

with possible negative mole numbers result for the six-component mixture (3.43). This example is relevant if a mole number reduction occurs
in the reactive system via a triple body collision process (which takes
place, for example, at high combustion chamber pressures). In this case
of restricted equilibrium, the order of the stoichiometric matrix $\underline{\nu}$ is
smaller than $\tilde{R} = K - E$. As Strehlow shows, using a representative set
of chemical equations for binary exchange reactions (1984, pp.102-103),
one obtains the equilibrium reaction scheme

$$H + H_2O \;\rightleftarrows\; \tfrac{3}{2}\, H_2 + \tfrac{1}{2}\, O_2$$

(3.50) $\qquad O + H_2O \;\rightleftarrows\; H_2 + O_2$

$$OH \;\rightleftarrows\; \tfrac{1}{2}\, H_2 + \tfrac{1}{2}\, O_2 \qquad\qquad ,$$

decisive for establishing the stoichiometric matrix $\underset{\sim}{\nu}$. The scheme now consists of only three reaction equations ($\tilde{R}$ = 3).

In each stage of the sequence of states, equations (1.25) and (1.26) are valid (at steady mass flow rates) for establishing the mole numbers in the reference state.

For this particular problem, then, scheme (3.45) is assigned to the re-active system (3.43) being discussed. The stoichiometric coefficients ν_{kr} can then be taken from the following matrix:

(3.51)

$$
\nu_{kr} =
\begin{array}{cccccc}
O & H & OH & H_2 & O_2 & H_2O \\
\end{array}
\begin{array}{c}
\leftarrow k = 1(1)6 \\
\downarrow r
\end{array}
$$

$$
\nu_{kr} = \left|
\begin{array}{cccccc}
1 & 1 & -1 & 0 & 0 & 0 \\
0 & 2 & 0 & -1 & 0 & 0 \\
2 & 0 & 0 & 0 & -1 & 0 \\
1 & 2 & 0 & 0 & 0 & -1
\end{array}
\right|
\begin{array}{c}
1 \\
2 \\
3 \\
4
\end{array} \;.
$$

Using this matrix and equations (1.24) to (1.26), a correct calculation of the speed of sound a of the reactive ideal gas mixture can be made.

It should be pointed out that the use of another coefficient scheme could produce substantial errors in the main operating data.

3.5 Nozzle Differential Equation; Mass Flow Eigenvalue

The correct determination of the properties of state in 'combustor' state $\underline{C}$ requires measures which cannot be reasoned strictly enough in classical gas dynamics. The solution - knowledge of all four properties $p_{\underline{C}}$, $\rho_{\underline{C}}$, $T_{\underline{C}}$, $v_{\underline{C}}$, and all $x_{k,\underline{C}}$ - is dependent on the given initial pro-perties (for example, $p_{\underline{F}}$, $h_{\underline{F}}$ and O/F) and on the mass flow rate $\dot{m}$ of the specified gas mixture of K components. It should be mentioned that the Lagrange multiplier θ , although it belongs to the solution, is on-ly of interest insofar as a value $\theta \neq 0$ is noticed in a displacement of the properties of state in $\underline{C}$ – differing from the NASA-Lewis Code solution, with $\theta \equiv 0$ corresponding to the minimization of the free enthalpy $\hat{G}(T, p, X_k)$.

The calculations for all properties of state in nozzle throat state $\underline{I}$ are rather sophisticated. The MM employs a typical thermodynamic constellation. The resulting solution is based on a highly abstract (considering the real nozzle reactive flows involved) argumentation. Thus a somewhat broader presentation of the line of argumentation is certainly more helpful than a short explanation. The whole discussion can be divided into three sections, each of which will be fully discussed below.

First of all, one should be reminded that the fixation of the properties of state in $\underline{I}$ for a rather artificial model fluid is explained in Section 3.3 of Part I as a commentary to equation (I-3.18). By characterizing this state with Mach number condition $M_{\underline{I}} \equiv 1$ – equation (3.21) and Section 2.2 in Part I – and given the parameter $\dot{m}$, the values of $p_{\underline{I}}$, $\rho_{\underline{I}}$, $T_{\underline{I}}$, $v_{\underline{I}} = a_{\underline{I}}$ and $X_{k,\underline{I}}$ are theoretically established. This situation is reflected by the following scheme $\cap_{\underline{I}}$:

$$
\begin{array}{ll}
(3.52) & \underline{C}:\ \hat{h}_{\underline{C}}^{tot}(\ldots|\dot{m}) = \hat{h}_{\underline{I}}(T,\ n) + \frac{1}{2}\left[\frac{RT}{\psi A}\right]_{\underline{I}}^{2} p_{\underline{I},i}^{-2} \;\longrightarrow \\[4mm]
& \qquad\qquad \longrightarrow\ \hat{T}_{\underline{I}}(p_{\underline{I},i}|\dot{m}) \quad \longrightarrow \\[3mm]
(1.15) & \qquad\qquad\longrightarrow\ \left[\begin{array}{l} \rho_{\underline{I}} = (\psi p_i/RT)_{\underline{I}} \\[2mm] a_{\underline{I}} = \sqrt{(\gamma RT/\psi)_{\underline{I}}} \\[2mm] v_{\underline{I}} = \dot{m}/(\rho A)_{\underline{I}} \end{array}\right] \longrightarrow \\
(3.40) & \\
(3.53) & \\[4mm]
(1.33.2) & \longrightarrow\ \left|\dfrac{v^2 - a^2}{v^2}\right|_{\underline{I}} \overset{?}{\leqslant} \epsilon_{\underline{I}} \xrightarrow{\ no\ } p_i + \Delta p_i \to p_i \to \underline{C} \\[2mm]
& \qquad\qquad\qquad\qquad\ \downarrow yes \\[2mm]
& \qquad\qquad\qquad\qquad\ p_i = p_{\underline{I}}
\end{array}
$$

The left side of equation (3.52) is the total mixture enthalpy given in $\underline{C}$; it is a functional of $\dot{m}$.

This iteration loop n_I (with the index i standing for the number of iterations which, for clarity, are only marked by the thermodynamic pressure p) makes the following iterative calculation possible: using equation (3.52) and the intermediate results for ρ, a and v as well as interrogating with equation (1.33.2) (for example, with $\epsilon_I \approx 10^{-4}$) and given p, one can obtain the temperature T in $\underline{I}$ and then all other properties. If a solution does in fact exist, then it pertains to the freely preselected mass flow parameter $\dot{m}$. Noteworthy enough, to find this solution one does not need any information on the nozzle contour along the axial coordinate.

Studying the set of equations above, it becomes apparent why the steady mass flow rate is treated as a free variable operating parameter in the NASA calculations. It differs from the respective scheme in the NASA-Lewis Code only in that its entropy condition (1.42) is not explicitly taken into account. As explained in Section 3.2, equation (3.52) is valid only for an inviscid change of state, so that the entropy condition ds = 0 is already, strictly speaking, taken into account. Using equation (1.32), S. Gordon and B.J. McBride achieve a pseudocorrection of the basic error in the iteration loop n_I : this error is discussed below as the **second** link in the chain of argumentation.

The error: the iteration scheme n_I for chemically reacting systems cannot actually be evaluated. The reason for this was discussed extensively in Chapter 1 and expressed, for example, in the formulation of equation (1.42). For changes in the composition of the gas mixture, one needs additional informations in order to register the change of state. If such processes occur in chemical equilibrium, calculation of this equilibrium must be made under consideration of the governing constraints. Section 3.2 offers a pertinent example of just such a calculation, i.e. for the consistent determination of the mole fractions $x_{k,\underline{C}}$, k = 1(1)K , in state $\underline{C}$. It is evident, however, that a suitable calculation scheme is not known for the equilibrium mole fractions $x_{k,\underline{I}}$ in $\underline{I}$, and for conditions which actually predominate during the sequence of states $\underline{C}$, $\underline{I}$ and $\underline{E}$. **The performance procedure of the NASA-Lewis Code always ignores this basic problem of process realization.**

Its calculation of the mole fractions according to the $\min\hat{G}(T,p,\mathbf{n})$-method for each actual paired data of T and p, is a theoretically crude approximation untenable in view of the strongly decreasing gas temperatures and pressures along the Laval nozzle. Such approximations naturally are incalculable risks in a situation of stringent standards.

The solution to this problem lies in a thermodynamic peculiarity: whenever complete information is available for two states, one may choose an appropriate (easily calculable) **substitute process** for the linkage between the two states. This possibility plays an especially important role in flow processes: they can be described with the Alternative Theory. That is, they need not be described with a thermostatic Gibbs-function, but rather (in the case, for example, of a flowing ideal polynary single-phase system) with a GFE in the form of equation (3.8). In such a case, the energy form of motion is also included.

Scheme $\cap_I$ naturally fulfills all the basic conditions for the use of a substitute process, as long as one assumes that the correct equilibrium mole fractions for state $\underline{I}$ in the first right hand term of equation (3.52) are involved; they are needed there for the mean molar mass ψ, and above all for the mole number vector $\mathbf{n}$ in the function $\hat{h}(T,\mathbf{n})$.
In this case, one can construct an abstract 'path' (based on the known values in state $\underline{C}$) which allows any number of states between $\underline{C}$ and $\underline{I}$ to be passed through until the final correct state data are reached in $\underline{I}$. This procedure is illustrated in the following <u>Figure 3.2</u>. Using the data ensemble 1 in state $\underline{C}$ (preliminary marking: $\underline{C}_1$) as the starting point and with a given parameter value $\dot{m}_1$, the mathematical mapping of this 'path' described above establishes pressure and temperature data along the dotted lines between states $\underline{C}$ and $\underline{I}$. Of course this mathematical form can be only a differential equation: hereafter it will be called simply the **'nozzle differential equation'**.

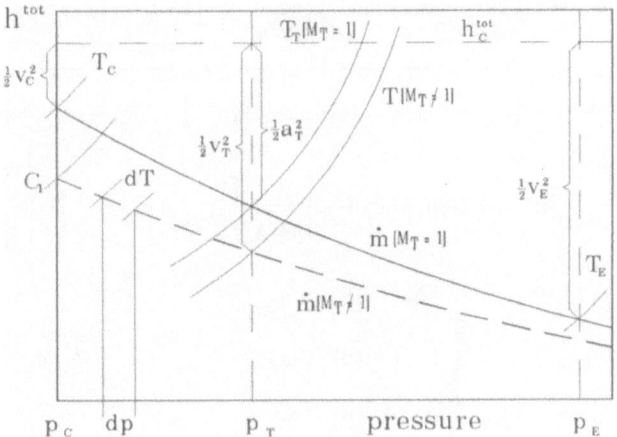

Figure 3.2: Construction of the Substitute Process

It is obvious that the solution $\hat{T}(p|\dot{m}_1)$ of this differential equation between the given paired data T and p in the states $\underline{C}$ and $\underline{T}$ has nothing to do with the real temperature and pressure distributions in the convergent part of a Laval nozzle, since the construction of the substitute process ignores (for simplicity's sake) the contributions of the gas velocity to the energy conversions. For this solution parametrisized by $\dot{m}_1$, virtual states **between** $\underline{C}$ and $\underline{T}$ are not submitted to an unchangeable total enthalpy requirement (as necessary **in** $\underline{C}$ and $\underline{T}$).

The purpose of the substitute process is solely to give such a **solution** Λ of the 'nozzle differential equation'

$$(3.54) \qquad \Lambda_{\underline{C} \to \underline{T}} := \hat{\Lambda}[T, p, n(T,p|minG(n))|\dot{m}] \equiv 0$$

for the chemically-reacting gas mixture, in which the elements n_{k} (or x_{k}) of the mole number vector n can be determined through the $min\hat{G}(n)$ equilibrium calculation using, for example, the AFC method.

The mole numbers n_{k} are allowed to change infinitesimally depending on the respective pair T and p. For a differential equation in these variables of state T and p, such a possibility can be realized in every

differential increment. Thus it becomes clear with which iteration scheme $\cap_{\Lambda,\underline{I}}$ the previously-mentioned scheme $\cap_{\underline{I}}$ can be replaced:

$$\underline{C}: \quad \Lambda \equiv 0 \quad \longrightarrow \quad \hat{p}_{\underline{I}}(T|\dot{m}_i) \quad \text{for } T = T_{\underline{I}} \quad \downarrow \qquad \qquad \Lambda \equiv 0$$

$$\longrightarrow h_{\underline{C}}^{tot}(\dots|\dot{m}) - \hat{h}[T,\hat{n}(\Lambda)]_{\underline{I}} - \frac{1}{2}\left[\frac{R\,T}{A\hat{\psi}(\Lambda)}\right]_{\underline{I}}^2 [\hat{p}_{\underline{I}}(T|\dot{m}_i)]^{-2} \overset{?}{\gtreqless} 0 \qquad \begin{array}{c} \uparrow \text{ no} \\ \\ \downarrow \text{ yes} \end{array}$$

$$\downarrow \left\{ \begin{array}{l} \text{Eq.(1.15)} : \rho_{\underline{I}} \\ \text{Eq.(3.40)} : a_{\underline{I}} \\ \text{Eq.(3.53)} : v_{\underline{I}} \end{array} \right\} \qquad \longleftarrow \quad \boxed{T_{\underline{I}}, p_{\underline{I}}}$$

$$\longrightarrow \left|\frac{v^2 - a^2}{v^2}\right|_{\underline{I}} \overset{?}{\ll} \epsilon_{\underline{I}} \begin{array}{c} \xrightarrow{\text{ no }} \\ \downarrow \text{yes} \end{array} \dot{m}_i + \Delta\dot{m}_i \rightarrow \dot{m}_i \quad \longrightarrow \quad \underline{F} \quad \overset{Eq.(3.18)}{\longrightarrow} \quad \underline{C}$$

$$\boxed{\dot{m}_{ev}}$$

This scheme begins with the solution $\Lambda \equiv 0$ (i.e. with an appropriate relationship between pressure and temperature in $\underline{C}$, which is functionally dependent on the chosen mass flow rate); the abbreviation $\hat{n}(\Lambda)$ means the mole number vector with components as functionals of temperature and pressure to be obtained from solutions Λ. The integration path runs normally from state $\underline{C}$ up to state $\underline{I}$.

The scheme illustrates a significant aspect of the MM: the iteration (index i) yields the **steady mass flow rate as eigenvalue** $\dot{m}_{ev}$, as described by K.N.C. Bray (in WEGENER (Ed.) 1970, p.82). Accordingly, the Lagrange multiplier θ also gains significance: in the iterative determination of this eigenvalue with scheme $\cap_{\Lambda,\underline{I}}$, θ regulates respectively the adjustment of the properties of state in state $\underline{C}$ for such an integration path which missed the state $\underline{I}$. This path leads from $\cap_{\Lambda,\underline{I}}$ back to initial state $\underline{F}$ and initiates the activation of the calculation scheme documented in equation (3.18). As one can see in a comparison with the results from Chapter 3 of Part I, θ takes over the function of parameter $\dot{m}$, inherent in correctly-solved classic gas dynamics problems.

Referring to Figure 3.2, one notes that the solution in the state space (not along the Laval nozzle!) is reached in principle because the mass flow rate-eigenvalue $\dot{m}_{ev}$ deviating from initial value $\dot{m}_1$ is now attributed to the properties of state in state $\underline{C}$. From these, the 'nozzle differential equation' can establish correct pressure and temperature data in state $\underline{T}$, and additionally, the sought mole fractions in chemical equilibrium as well.

Naturally the question arises as to how this 'nozzle differential equation' can be obtained. The answer is the **third** link in the chain of argumentation. It is relatively easy to obtain, since this differential equation is a "by-product" in the derivation of the speed of sound in chemically-reacting fluids.

The primary idea of the substitute process is related to an isentropic change in pressure and temperature of the investigated combustion gases at a hypothetical state of rest (see equation (2.8) in the final section of Appendix 2). This change occurs during the transition from the known data in state $\underline{C}$ to the equally known data in the final state $\underline{T}$. Calculation of the accompanying equilibrium mole fractions can be made for each differential increment (under isobaric-isothermal constraints) using the AFC method discussed in Section 1.2.

The sought differential equation for the substitute process is already present in equation (3.39). If one uses

(3.55) $\qquad\qquad y := \ell n\ p \quad ; \quad x := \ell n\ T \quad ,$

then the equation can be more compactly expressed

(3.56)
$$\frac{dy}{dx} = \frac{c_{p,\xi}^m/R + \Gamma_T}{1 - \Gamma_p}$$

Due to the strong dependence of the right-hand properties on x and y, the equation is non-linear.

With equation (3.56) one obtains a precise solution to the problem of calculating the properties of state in state $\underline{T}$ without knowing more than three cross section areas of the nozzle geometry. This 'nozzle differential equation' also plays a decisive role in establishing the properties of state in the final cross section area $\underline{E}$, as well as in treating the problem of cooling a rocket engine.

In the case of the change of state from $\underline{C}$ to $\underline{I}$, the differential equation includes a kind of **inverse boundary value-problem** in which the function data for two end points are not, as usual, fixed, and a function is sought which includes these two salient points in its data under certain constraints. The 'nozzle differential equation' is given, yet the coordinates for the end points are not definitively established and have to be fixed using a parameter variation. Parameter here is the steady mass flow rate $\dot{m}$; it has an eigenvalue of $\dot{m}_{ev}$, which is characteristic for the ICP.

3.6 Nozzle Exit State: Problems of Numerical Computation

The NASA-Lewis Code's theoretically inadequate iteration procedure for establishing the nozzle exit pressure $p_{\underline{E}}$ was extensively discussed in Section 1.5. Equation (1.43) is characteristic for this numerical matching manipulation, which establishes pressure $p_{\underline{E}}$ for a chosen nozzle final cross section area $A_{\underline{E}}$, using a given set {....} of operating parameters in state $\underline{F}$ such as $\{p_{\underline{F}}, h_{\underline{F}}$ and $O/F\}$, as target function.

The same problem implies a dilemma when solved by the MM: although it is not particularly severe, it is not without theoretical inferences:

- on the one hand, $p_{\underline{E}}$ is a relevant target function for calculating the resulting thrust in design studies for various nozzle final cross section areas and for varying sets {....} of initial data ('Calculation I');

- on the other hand, the MM is based on the fundamental condition of reversible changes of state throughout the sequence of states. Thus it requires retaining condition $p_{\underline{E}} \geq p_U$ (see Section 2.2 of Part I) – ('Calculation II').

This dilemma has a technically interesting and physically satisfactory solution: if one introduces the condition

$$(3.57) \qquad\qquad p_{\underline{E}} \overset{!}{\geq} p_U$$

for realization of the process, the nozzle exit pressure p_E is no longer a calculable property of state, but at least equal to the surrounding fluid pressure p_U . One now has the possibility of altering either
the cross section area at the nozzle exit (or at another characteristic
location in the nozzle's configuration), or one of the available operating parameters in $\underline{F}$ so that condition (3.57) is at least fulfilled.
The parameters p_E and O/F are the theoretically preferable choices.

Since the surrounding fluid pressure p_U (or the thrust, which changes
linearly with p_U - see BARRèRE et al. 1960, p.93) can generally be chosen so that it is, for example, suited for the typification of a **flight
mission** $\hat{z}(t)$, one can use $\hat{p}_U(z|t)$ to obtain the optimal relationship
between the **profile of the surrounding fluid pressure** ($\equiv$ ambient pressure)

(i) $\hat{p}_U(z|t)$ and the feed pressure p_E

(ii) $\hat{p}_U(z|t)$ and the mass flow ratio O/F

when other conditions are constant. It is evident, especially with regard to reduction of fuel consumption in an aircraft, that such a possibility offers an attractive technological challenge to the manufac-
turers of future feed systems! The MM is particularly notable because
its thermodynamic optimization systematically takes quantitative mission planning into consideration from the very beginning. This remarkable advantage is especially important for the forecasted expensive
design work needed for the next generation of hypersonic aircraft.

As a rule, such mission profiles are currently not taken into consideration in the early planning stages for high-enthalpy engines. Thus it
is less than appropriate to give random p_U-data only in order to formally satisfy condition (3.57). It suffices to assume that the calculated nozzle exit pressure conforms to equation (3.57) in 'Calculation
I'. In individual cases, the influence of the supersonic nozzle's geometric form should be included (see BARRèRE et al. 1960, pp.69 f).

'Calculation II' is carried out virtually analog to the iteration sche-
me $\cap_{\Lambda,\underline{I}}$ for establishing the properties of state in the nozzle throat
cross section. Here it is again decisive that the states $\underline{I}$ and $\underline{E}$ are

known 'in principle'. Through the calculations of the mass flow eigen-value documented in the previous section, the properties of state in $\underline{I}$ at the known geometric properties $A_{\underline{I}}$ and $\alpha := A_{\underline{C}}/A_{\underline{I}}$, as well as the operating parameters in $\underline{F}$, are firmly assigned. Four equations are available for the four properties of state in $\underline{E}$: the continuity equation for $v_{\underline{E}}$ at the known $\dot{m}$-eigenvalue, the thermal equation of state for $\rho_{\underline{E}}$, the First Law for reversible open systems for $T_{\underline{E}}$, and condition (3.57) for $p_{\underline{E}}$. It is assumed that informations for the correct calculation of the mole fractions $\chi_{k,\underline{E}}$, $k = 1(1)K$ in chemical equilibrium are available. For the reasons given in Section 3.5, there is a possibility for the substitute process being realized between states $\underline{I}$ and $\underline{E}$. It is described with the same 'nozzle differential equation' (3.56), whereby its **numerical solution** $\Lambda_{\underline{I}\to\underline{E}}$

$$(3.58) \qquad \Lambda_{\underline{I}\to\underline{E}} := \hat{\Lambda}[T,p,\hat{n}(T,p|\min\hat{G}(n))]|p_{\underline{F}}] \equiv 0 \qquad\qquad 58$$

can be utilized for the necessary iteration scheme $\cap_{\Lambda,\underline{E}}$:

$$\underline{I}: \qquad \text{Eq.}(3.57): p_{\underline{E}} \overset{!}{=} p_U \to$$

$$\to h_{\underline{F}}^{tot} - \hat{h}[T,n(\Lambda)]_{\underline{E}} - \frac{1}{2}\left[\frac{R\,T}{A\psi(\Lambda)}\right]_{\underline{E}}^{2}[\hat{p}_{\underline{E}}(T|p_{\underline{F},i})]^{-2} \equiv 0 \to T_{\underline{E},i}$$

$$\to \left[\begin{array}{l}\text{Eq.}(1.15): \rho_{\underline{E}} \\[2mm] \text{Eq.}(3.53): v_{\underline{E}}\end{array}\right] \to$$

$$\to \Lambda_{\underline{I}\to\underline{E}} \equiv 0 \to \hat{T}_{\underline{E}}(p_{\underline{E}}|p_{\underline{F},i}) \hspace{3cm} \underline{I}$$

$$\to |\hat{T}_{\underline{E}}(p_{\underline{E}}|p_{\underline{F},i}) - T_{\underline{E},i}| \overset{?}{<} \epsilon_{\underline{E}} \xrightarrow{\text{no}} p_{\underline{F},i}+\Delta p_{\underline{F}} \to p_{\underline{F},i}$$

$$\downarrow \text{yes}$$

$$\boxed{T_{\underline{E}}, n_{\underline{E}}} \hspace{2cm} ;$$

The iteration index i is once again fixed — for transparency — to both the iteration variables p_F and[59] (to distinguish it from the temperature T_E derived from solution Λ at the given pressure p_E) the temperature obtained from the total enthalpy equation. At the given final cross section area A_E, the scheme $\cap_{\Lambda, E}$ produces, only via an extensive iteration loop, the sought property of state in $\underline{E}$. One must consider that any change of an initial parameter such as p_F activates all previously-given iteration schemes. In such a case, for example, both the Lagrange multiplier θ and the mass flow eigenvalue are also changed!

'Calculation II' is thus relatively time-consuming and, in a comprehensive design study including calculations of structural masses, should normally be put aside in favor of 'Calculation I'. This simplified iteration procedure uses, with the exception of condition (3.57), all the given relationships. They are no longer a functional of p_F (or O/F) but rather a functional of the final cross section area A_E. With the known data in state $\underline{I}$ one obtains the properties of state T_E, p_E and the mole number vector n_E from a simultaneous iteration of the (total) enthalpy equation and the solution Λ of the nozzle differential equation. Properties v_E and ρ_E follow immediately from the continuity equation and thermal equation of state. One thus obtains a table in which the **property of state vector** for the state $\underline{E}$ is entered above A_E.

If one defines the parameter

$$(2.2\text{-}I) \qquad \epsilon := A_E / A_I$$

as a number related to the cross section area A_I in the nozzle throat, it is possible (with fixed value for A_I) to obtain the respective property of the state vector for $\underline{E}$ for each given ϵ-value through interpolation of the available table. Since the tabular presentations of such computations are applicable for the NASA-Lewis Code, they will also be used in this study. This presentation contains a complete summary of all relevant operating parameters, area informations, optimization parameters such as θ and $\dot{m}$, as well as the properties of state for all states $\underline{F}$, $\underline{C}$, $\underline{I}$ and $\underline{E}$ (the latter quantities dependent on the area ratio ϵ). In addition, some data in the table is related to conditions which must be met, such as s = constant, h^{tot} = constant and M_I = 1. Finally, the table also includes the actual target properties of the MM: the specific (second) impulse as well as the thrust of the investigated rocket engine (for vacuum conditions and for the obtained nozzle exit pressure), calculated according to the definitions presented in Part I.

In order to avoid misunderstandings, it should be stressed that the properties of the fuel-oxidizer mixture in state $\underline{F}$ refer to a non-ignited fluid state. The 'flame state' originally introduced in the tabular Lewis Code presentation is now dismissed.

This concludes the basic version of the Munich Method. The results of this procedure are recommended as a reference. As a comparative process, the developed physical model cannot **provably** be improved.

Therein lies the substantial advantage of the MM over all competitive procedures! The calculated results do not contain physical or methodological defects provided the AT concept is accepted as a baseline. Obviously there are possibilities of both mathematical errors – as a result of programming gaps, erroneous mathematical algorithms and computer limitations – and uncertainties due to the use of inadmissable material data. Continuous program maintenance is thus indispensible!

The basic version of the Munich Method offers notable possibilities for both methodological and practical (engine-oriented) amplifications. Further work in the latter field requires a detailed prior analysis of the typical features and peculiarities of each high-performance propulsion system to be investigated. This preliminary work has been accomplished by P.Kramer in an excellent and comprehensive study. His meticulous investigation covers everything from seemingly simple ram-jet engines to the most advanced turbo scram-jets (KRAMER 1987). The correct application of the most important thermofluiddynamic processes to such propulsion systems, using procedures of the Munich Method, is extraordinarily complex and involves unforeseeably numerical risks; it must be investigated in separate studies. In this context, G. Kappler has recently presented some remarkable arguments in favor of the technological research for a new generation of aero-engines; future studies would obviously be rewarding, particularly with respect to trans- and hypersonic propulsion systems (see KAPPLER 1988, pp.24 f).

A general methodological expansion of the MM, on the other hand, is considerably simpler and also preferable: it would naturally be of help for subsequent investigations related to applications. Two especially interesting investigative goals should be mentioned at this point.

The first is the expansion of the MM to include systems in which the combustion enthalpy h_F of the fuel-oxidizer mixture in state $\underline{F}$ is not much higher than the specific kinetic energy $\frac{1}{2}v_F^2$ of the mixture in this state. Although an analysis of the consequences of such a change in the initial data is exceedingly complex for a number of reasons, such work would undoubtedly be valuable for the development of high-performance propulsion systems with trans- or hypersonic combustion.

The second investigative goal is basically more important: a study of the cooling requirements for such high-enthalpy nozzle flows. Even the concept of this particular theme is extremely difficult: since, in principle, the energy transfer processes are irreversible at finite temperature differences, the concept of the ideal comparative process (ICP) has to be modified. Above all, its contents must be reconsidered. The final section of this chapter is intended as a step in this direction: it deals with some basic aspects of **regenerative cooling** of an exemplary rocket engine (the SSME type used for the space shuttles) and its thermodynamic modeling. Relevant numerical results are presented and discussed. Completion and proof of a problem-oriented, profound theory of regenerative rocket engine cooling, however, must be the subject of a separate study.

As a conclusion to this section, the basic version of the MM is summarized in the form of a structural diagram. In addition, a number of problems encountered in the numerical application of the theory are outlined. Since at the moment the MM can be worked out with an adequate numerical code only for a H_2-O_2-combustion in a conventional rocket engine, the structural scheme is specifically tailored to a computer program named 'CHOPER'.

The main scheme documented in Figure 3.3 shows a formal structure which essentially is self-explanatory. It consists of three characteristic iteration loops oriented to the typical sequence of states in the ICP.

These loops were presented in sections 3.2, 3.5 and 3.6 as separate schemes for the changes of state between states $\underline{F}$ and $\underline{C}$, then $\underline{C}$ and $\underline{I}$, and ultimately $\underline{I}$ and $\underline{E}$. The marked places in which the iteration variables for these schemes – the parameter θ, $\dot{m}_{ev}$ as well as $p_{\underline{F}}$ (or O/F) or $A_{\underline{E}}$ respectively – can be used in the program, are clearly recognizable. If one wants to carry out 'Calculation II', the program contains four intertwined main iteration loops. Additional sub-iteration cycles are also required for calculating the local equilibrium composition, the solution of the (total) enthalpy equation, as well as for the integration of the 'nozzle differential equation'. The basic mathematical relationships are, without exception, strongly non-linear and require relatively considerable computing time. An optimization of the computer program is therefore unavoidable in view of extensive design investigations with systematic variations of the numerous operating parameters. This problem was encountered during the development of CHOPER's present version. Although it was originally planned to work with more accurate thermodynamic functions rather than with NASA's polynomial set, this attempt had to be abandoned: the radical increase in computing time made the idea impractical. This decision is problematic since, as NEBENDAHL et al. (1987) and WILHELMI (1987) have shown, the NASA polynomials (based on the JANAF tables) are not reliable above 3000°C: they can no longer be extrapolated with sufficient accuracy. This difficulty does not occur with the method based on applied statistical thermodynamics, thus giving the method decisive advantages (see LUCAS 1986, pp. 1 & 21) – but only disregarding time consumption.

The problem can be resolved only by substituting the 25-year-old NASA polynomials with more modern instruments of numerical mathematics and subsequently checking the new results using statistical thermodynamics. This check should also include the reactive system (3.43) in the domain defined within bounds

(3.59) $500 \leq T[K] < 4000$; $10 < p[bar] \leq 1000$

of common properties of state. Effects of the intermolecular interaction (of the Van der Waals type) should be considered with regard to the thermal and caloric equations of state.
A solution for such problems has not yet been found but is urgently needed simply because of the exemplary significance of H_2-O_2-combustion

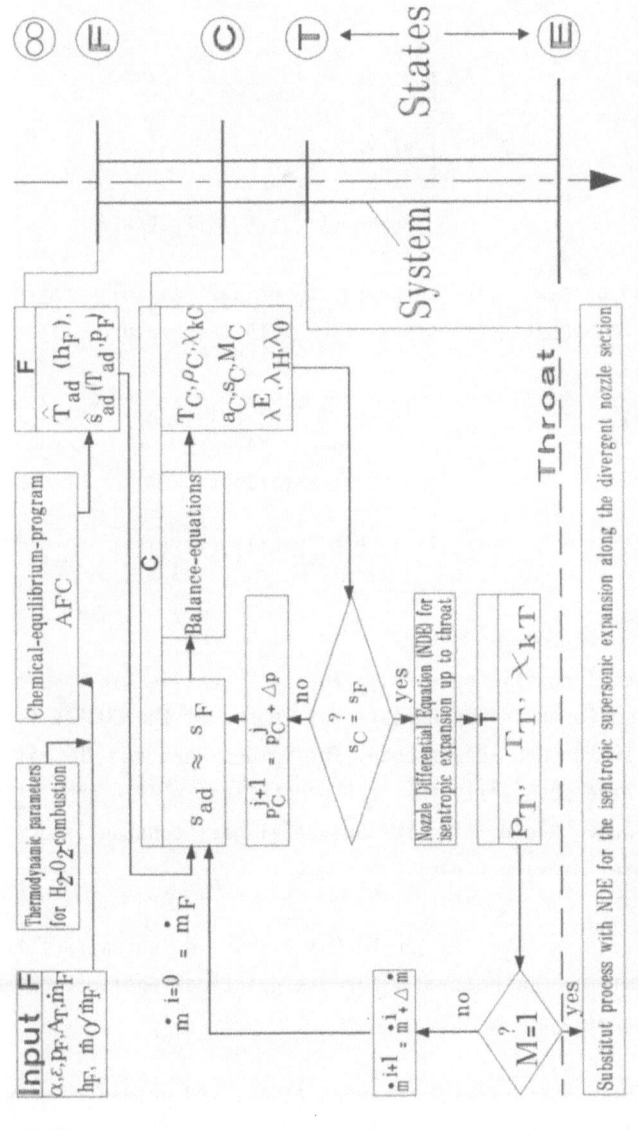

Figure 3.3: Scheme of the Computerprogram CHOPER (Combustion Hydrogen Oxygen Program Engine Rocket)

for similar problems in other important reactive systems. At this point
one can only make conjectures about which rational (i.e.reliably inter-
and extrapolational) equations, Spline functions or data banks can be
used to substitute for the presently-used NASA polynomials for the molar enthalpy h^m or the molar entropy s^m of the chemically reacting mixture. These polynomials have the form (compare equations (1.22) and
(1.30)):

(3.60) $$h^m := \sum_k h^m_k \, X_k \quad ; \quad k = 1(1)K \quad ;$$

(3.61) $$s^m := \sum_k s^m_k \, X_k$$

$$s^m_k = \hat{s}'_k(T) - R \, \ln[X_k \, p/p^\varnothing] \quad ,$$

whereby the partial molar enthalpy h^m_k or partial molar entropy $\hat{s}'_k$ for
ideal gas components are approximated with temperature polynomials (see
Appendix 2):

(3.62) $$h^m_k = R \sum_j j^{-1} \beta_{j,k} \, T^j + \beta_{6,k}$$

$$j := 1(1)5$$

(3.63) $$\hat{s}'_k(T) := R[\beta_{1,k} \ln(T)] + \sum_j (j-1)^{-1} \beta_{j,k} \, T^{j-1} + \beta_{7,k}$$

$$j := 2(1)5 \quad .$$

The polynomial coefficients $\beta_{j,k}$ for the reactive six-component system
(3.43) are stored in appropriate data banks of the CHOPER Code as well
as listed in the NASA-Lewis Code. The values need not be listed here.
They each have such unity that the properties h^m_k/RT and $\hat{s}'_k(T)/R$ are
dimensionless; R signifies the universal gas constant and $p^\varnothing$ a standard pressure (usually 1 bar).

Two peculiarities noted in the LH-LOX mixture do not affect the rocket
theory, but may be relevant for its practical application at today's
level of knowledge:

(1) the nozzle differential equation (3.56) is primarily determined by
    the molar enthalpies h^m_k of the gas mixture's six components. Its
    solution in the respective form of labeled triple T_i, p_i and $X^{equ}_{k,i}$,
    $i = 0(1)I$, apparently reproduces the required unchanged specific

entropy $s_i = s_0$ of the initial value (i = 0) from equations (3.61) and (3.63). A numerical agreement within given bounds is only taken for granted if the coefficients $\beta_{6,k}$ and $\beta_{7,k}$ for all components are mutually consistent. Based on experience, this condition for the NASA polynomials is not sufficiently met. For comparative reasons only, the original β-constants were retained in the current version of the CHOPER Code. Future modifications of this computer program need improved thermodynamic material functions.

(2) As explained in Appendix 2, there is currently no possibility of obtaining the necessary mixture specific entropy s_F in **fluid** state from the corresponding data for LOX and LH.

The approximation $s_F \approx s_{ad}$ (i.e. the unknown entropy value to be approached via the solution for $\alpha \to \infty$ according to the AFC method) is only justifiable for the initial calculation. The application of s_{ad} under the constraint $1 < \alpha < \infty$ is as a rule physically not recommendable.

With the hypothetical value of s_{ad} and the solution of the algebraic equations (3.18), one obtains at least estimates for the data on the gas mixture's specific kinetic energy $\frac{1}{2}v^2$ and the local pressure p expected in state $\underline{C}$. They may be used in all cases of practical importance for estimating the fluid entropy s_F: with the AFC method, one can calculate, for example, a fictive combustion chamber temperature T_B and gas concentrations $X_{k,B}$ using the specific enthalpy $h_B := h_F - \frac{1}{2}v_C^2$. Together with a weighted mean value from both the pressures p_F and p_C (which takes the pressure drop into account), one can immediately get close to s_F with equations (3.61) and (3.63). Observing the exposition in Appendix 2, one may establish the result as the obligatory ICP specific entropy assumed as unchangeable along the sequence of states.

The current version of the computer program CHOPER for the H_2-O_2-combustion is fully documented and described elsewhere (LIPPIG 1987). It was programmed according to the MM by W. Düster and subsequently tested. The code consists of approximately 5000 program lines. The program's modular structure covers the following tape and manual requirements:

(i)   representation of the theory (MM)

(ii) transparency of the source code

(iii) quick convergence

(iv) portability.

The CHOPER Code was developed on a Uni-Sys A 15 computer. The sub-pro-
grams were successfully tested on various PCs.[60]The capability of both
the MM and the computer program have been subsequently checked and verified for numerous important examples. These will be discussed in Chapter 4.

3.7 Regenerative Cooling: Attempts towards a Modeling[61]

High-performance engines with H_2-O_2-combustion have to be efficiently
cooled. This requirement is particularly important for reusable systems
such as the main engines of the space shuttle (SSME).
A theoretical understanding of the cooling problem is especially important for establishing appropriate scaling procedures for liquid-fuel
rocket engines. L.G. Crocco was the first who presumably worked out
rules on similarity of regenerative cooling systems (see PENNER 1957,
pp.381 f).
Numerous well-tested **cooling techniques** exist for a variety of propuls-
sions, including rocket engines (see MIELKE 1986, pp.236-238). In liquid-propellant rockets, regenerative, circulating or sweat cooling
arrangements are the usual methods. Film cooling is the most favored
procedure.
Regenerative cooling is generally used with liquid-propellant rockets.
Either the oxidizer or the fuel, or both, can be employed as a cooling
agent circulating around the motor through a jacket or in a helicoidal
channel. This cooling method, however, is often subject to a serious
and widespread misunderstanding. The following quote concisely defines
the problem: "Combustion chamber and nozzle have been constructed with
double walls: the fuel is guided through this void to cool. In doing
so, it is heated up before it reaches the combustion chamber: we haven't lost any energy (regenerative cooling)." (RUPPE 1982, p.26). This
statement obviously implies that the cooling devices are free of any
losses. Yet this is only partially true! As previously mentioned, the

convective transfer of heat at a finite temperature difference is always irreversible; the entropy generation S_{gen} is (BEJAN 1982, p.36)

$$(3.64) \qquad S_{gen} = Q\left[T_{low}^{-1} - T_{high}^{-1}\right] \geq 0 \quad .$$

Q signifies the heat to be transferred between a high and a low temperature level. If the core temperature in the nozzle throat cross section, for example, lies 2000°C over the mean coolant temperature, one can expect enormous losses ultimately affecting the total balances of mass, momentum and energy. Some authors have emphasized the relevant connections between reaction and cooling rates (see PENNER 1957,p.309). Further discussion of these aspects will be offered in a future study.

Equation (3.64) offers a first indication of the conceptual difficulties involved in transforming an ICP to an appropriate, irrefutably dissipative comparative process with regenerative cooling. Reasonable approach would first deal with this exergy problem in greater detail. The concept of regenerative cooling itself has to be subjected to a thorough energetic analysis. The following discussion should thus be considered only provisional.

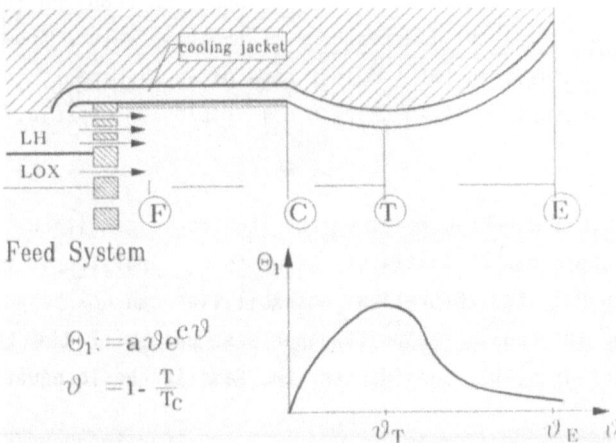

Figure 3.4: Modeling of Cooling; Geometrical Configuration

<u>Figure 3.4</u> illustrates a typical geometric constellation with a cooling channel running along the outer contour of the Laval nozzle and feeding into the LH main feed line or connecting with the injection system (see BARRèRE et al. 1960, pp.447 f).

In addition, the plot above shows the variation of property θ_1 along the coordinate θ; the latter is coupled to gas temperature T in such a way that θ is reduced to zero at $T_{\underline{C}}$. Property θ_1 is intended to represent the concept of **cooling capacity** q so that q is at a maximum in the nozzle throat region. Such a requirement naturally can be modified at will; it is chosen as an illustrative example rather than to offer practical instructions. To further simplify matters, the combustion chamber itself has been omitted from the cooling plot, although this by no means affects the fundamentals discussed here.

In order to determine the cooling capacity q, the modeling begins by abstracting the concrete nozzle contour and then re-examining the sequence of states, this time from $\underline{C}$ through $\underline{I}$ to $\underline{E}$. At this point, no use is made of the intermediate states.
In line with the MM a theoretically important fact is considered correctly in this model: the isentropic condition required for ICP-realization nullifies the independence of the temperature from the pressure (and vice versa); in the simplest case of an elementary ideal gas, the known isentropic relationship $p \simeq T^\kappa$ yields an especially instructive example.

In any other cooling modeling the isentropic condition is replaced by another appropriate constraint. Assuming a **'polytropic coupling'** between p and T, two theoretical possibilities can now be adopted for associating the cooling capacity q per mass unit with the thermodynamic properties of state. Consider the two familiar basic equations

(3.66) $\bar{d}q/T = ds$ (Carathéodory/Planck)

(3.65) $s_2 - s_1 - \int_1^2 T^{-1}\, dq \geq 0$ (Clausius)   ;

there is a subtle difference between the two terms!

The first euqation makes physical sense only as long as one interprets temperature T as an integrating factor; such an assumption is invalid for thermodynamic with more than two variables. The second equation is also disputable since the integral between the states 1 and 2 cannot be solved until the problem-oriented connection between q and T is known.

The situation immediately changes when one is allowed to assume a relationship between q and T as the only variable, as a consequence of the polytropic coupling.

Such a **'modeling of the cooling'** is motivated by the idea that energy is transferred to the surroundings between two neighboring gas states along the nozzle. This transferred energy q is coupled with a dimensionless function θ_* in such a way, that the desired variation of q can be obtained quantitatively. This is also true for the substitute process defined by the appropriate nozzle differential equation introduced by equation (3.56) for the ICP.
The definition

$$(3.67) \qquad q := q_{\underline{C}}\, \theta_* = q_{\underline{C}}\, \hat{\theta}_*(T\,|\,T_{\underline{C}},\ T_{\underline{I}},\ T_{\underline{E}})$$

$$:= q_{\underline{C}}[1 + \hat{\theta}_1(\theta\,|\,1,\ \theta_{\underline{I}},\ \theta_{\underline{E}})]$$

may be chosen in such a manner that the **modeling function** θ_* is dependent on the (non-dimensional) gas temperature θ, and is a functional of the temperatures in the states, between which the cooling process takes place. In the case presented here, the states $\underline{C}$, $\underline{E}$ and $\underline{I}$ have been chosen as references.

Within certain limits the cooling capacity q may be specified by given requirements for the rocket engine performance. For simplicity, q is first fixed at a maximum value with regard to θ at nozzle state $\underline{I}$. Other agreements are recommended for theoretical analysis.

In order to establish clearly the course of $\hat{q}(\theta)$, the following information is needed:

(i)   in order to quantify the property $q_{\underline{C}}$, which regulates the unit and domain of function $\hat{q}(\theta)$, the Clausius inequation for the reversible limit

$$(3.68) \qquad q_{\underline{C}} := T_{\underline{C}} \; \Delta s_{\underline{C} \to \underline{I} \to \underline{E}} \left[\int_0^{\theta_E} \frac{1}{1 - \theta} \; [\frac{d}{d\theta} \; \hat{\theta}_1(\theta)] \; d\theta \right]^{-1} \quad ,$$

is used, whereby the dimensional temperature $\theta := 1 - (T/T_{\underline{C}})$ is substituted for the gas temperature T . With the known reversible entropy changes $\Delta s_{\underline{C} \to \underline{I} \to \underline{E}}$ of the combustion gas mixture as well as the prescribed profile θ_1 (see Figure 3.4), one can calculate the 'amount of heat per mass unit' $q_{\underline{C}}$. The reference of $\hat{q}(\theta)$ to state $\underline{C}$ identifies $q_{\underline{C}}$ as the cooling capacity transferred here.

(ii) the free constants of the chosen profile $\hat{\theta}_1(\theta)$ (given in Figure 3.4) are elementarily determined by the specified temperature values in states $\underline{C}$, $\underline{I}$ and $\underline{E}$, easily calculated with the MM.

The cooling capacity q has to meet two essential requirements: prevention of the nozzle wall's destruction and minimal reduction of the rocket engine's thrust. The cooling modeling is thus based on a kind of disturbance-ansatz for the properties of state in the sequence of states. In a first step, the free constants of $\hat{\theta}_1(\theta)$ can be determined through $T_{\underline{C}}$, $T_{\underline{I}}$ and $T_{\underline{E}}$ of the isentropic solution obtained with the MM. In order to eliminate the incompatibilities arising in this procedure, it is possible to modify the basic version of the MM itself in a second approximation.

First equation (1.36) for the total enthalpy between $\underline{C}$ and $\underline{I}$, and $\underline{I}$ and $\underline{E}$ is modified as follows:

$$(3.69) \qquad q + h + \tfrac{1}{2}v^2 = h_F^C \quad ;$$

in this expression the role of the First Law as a mediator between system and surroundings is made especially clear. Now, not only is the cooling capacity (= amount of heat transferred from the system across its boundaries) q added, but also **a change of the total enthalpy in the initital state $\underline{F}$ appears in comparison with the reversible-adiabatic standard solution** obtained from the MM. Since part

$$(3.70) \qquad \dot{m}_c := \zeta \; \dot{m}_{LH}$$

of the total fuel mass flow rate $\dot{m}_{LH}$ is initially diverted for cooling, then warmed with q and finally remixed with the remaining fuel flow rate

$$(3.71) \qquad \dot{m}_{H_2} = (1 - \varsigma)\dot{m}_{LH} \quad ,$$

the specific enthalpy $h_{\underline{F}}$ of the LH-LOX mixture (calculated from equation (1.28) for the standard case) changes in the following manner; ς is the coolant share.

This external routing of the fuel mass flow rate is usually termed **'regenerative'** (see SUTTON 1986, p.203). The expression refers to the intention of returning the energy q per mass unit withdrawn via the coolant flow rate $\dot{m}_c$ to the combustion gas. Insofar as one can attribute the unavoidable losses - quantified in equation (3.64) - to aggregates outside of the 'nozzle cooling jackets', one may actually allow the validity of such a hypothesis if one also assumes a negligible pressure drop within the liquid coolant flow itself (see SUTTON 1986, p.204). In this case, application of the First Law to the cooling system via

$$(3.72) \qquad \dot{m}_c \; \Delta h_c = \int_0^{\theta_{\underline{E}}} \hat{q}(\theta) \; \dot{m}_{ev} \; d\theta$$

expresses the transferred energy. The difference of the specific enthalpy Δh_c is transmitted from the combustion gas with its steady mass flow rate $\dot{m}_{ev}$ between entrance and exit of the coolant flow. Using the definition (3.67) given above for the cooling capacity $\hat{q}(\theta)$ at steady mass flow rates, and observing the definitions

$$(3.73) \qquad \dot{m}_{ev} := \dot{m}_{LOX} + \dot{m}_{LH}$$

$$(3.74) \qquad O/F := \dot{m}_{LOX}/\dot{m}_{LH}$$

for the optimized total mass flow rate $\dot{m}_{ev}$ and the mass flow ratio O/F, one arrives at the relationship

$$(3.75) \qquad \Delta h_c = \frac{1}{\varsigma}(1 + \frac{O}{F}) \; q_{\underline{c}} \left[\theta_{\underline{E}} + c^{-2}a \; [e^{c\theta_{\underline{E}}}(c\theta_{\underline{E}} - 1) + 1] \right]$$

after integration. The free constants a and c are introduced by the exponential form chosen for θ_1 (see Figure 3.4). Apparently the enthalpy difference Δh_c is primarily dependent on parameter $q_{\underline{c}}$, defined by equation (3.68) and thus explicitly determined by the entropy difference $\Delta s_{\underline{C}\rightarrow\underline{I}\rightarrow\underline{E}}$.

To a certain extent one can freely choose this entropy difference occurring in the combustion gases. For initial design calculations, however, such a free selection is impractical. It is both easier and more
expedient to treat the enthalpy difference Δh_c as a selectable parameter. With given inlet data T and p of the coolant flow as well as with
a negligible pressure loss, the **desired** coolant temperature at the end
of the channel establishes this enthalpy difference Δh_c; the appropriate value of the refrigerant can be taken from standard tables (see NE-
BENDAHL et al. (1987) for supercritical hydrogen properties).

Thereafter it is practical to resolve equation (3.75) for $q_{\underline{C}}$ and use
the resulting expression in equation (3.68) to determine $\Delta s_{\underline{C} \to \underline{I} \to \underline{E}}$. One
then obtains the pragmatic relationship

$$(3.76) \qquad \Delta s_{\underline{C} \to \underline{I} \to \underline{E}} = - \beta \, T_{\underline{I}}^{-1} \Delta h_c$$

between the entropy difference in the nozzle flow and the enthalpy difference in the coolant flow. For this, the (dimensionless) '**cooling coefficient**'

$$(3.77) \qquad \beta := \frac{T_{\underline{I}}}{T_{\underline{C}}} \frac{\zeta}{1+O/F} \left[\int_0^{\theta_{\underline{E}}} (1 - \theta)^{-1} \left(\frac{d\theta_1}{d\theta}\right) d\theta \right] \left[c^{-2} a[(1 - c\theta_{\underline{E}})e^{c\theta_{\underline{E}}} - 1] - \theta_{\underline{E}} \right]^{-1}$$

is introduced. It is dependent on the factor ζ, the mass flow ratio
O/F, the temperatures in the states $\underline{C}$, $\underline{I}$ and $\underline{E}$, and on the chosen cooling profile θ_1.
The process parameter β can be easily computed numerically and broadly
characterizes the cooling process under the given circumstances.

As long as adequate experience with realistic modeling of the cooling
system is not available [62] , such a computation is not particularly
helpful. This is true regardless of the choice of the modeling function
θ_* (equation (3.67)). Even taking possible incompatibilities into account, it is preferable to study the cooling process in a case where
the cooling coefficient β equals one. For a definite modeling in such a
case, one can at least estimate the influence of the cooling process on
the most important data relative to the isentropic standard case. This
was done in the present study!

With the establishment of [63] $\beta \equiv 1$, and given Δh_c for the liquid [64] coolant, one can first calculate $\Delta s_{\underline{C} \to \underline{T} \to \underline{E}}$ and then $q_{\underline{C}}$ using equations (3.76) and (3.68). This calculation requires knowledge of the temperature data in the corresponding states. In order to approximate them it should be noted that the properties of state for the standard case (with h_F) are not at all practical as initial values. Of greater use is the temperature data for the **reference case**, which is defined as a reversible adiabatic sequence of states under the same conditions, yet with reference to the modified initial enthalpy $h_{\underline{F}}^C$ (equation (3.69)).

In order to establish this property of state typical for regenerative cooling, the total enthalpy needs to be balanced for the state $\underline{F}$

$$(3.78) \qquad \dot{m}_{ev} \, h_{\underline{F}}^C = \dot{m}_c \, h_c + \dot{m}_{H_2} \, h_{H_2} + \dot{m}_{LOX} \, h_{LOX} \quad .$$

Together with equations (3.70), (3.71) and (3.74) and the simple summations rules

$$(3.79) \qquad \omega_{LH} = (1 + O/F)^{-1} \quad ; \quad \omega_{LOX} = 1 - \omega_{LH}$$

between the mass flow ratio O/F and the mass fractions ω_{LH} and ω_{LOX} of the two components LH and LOX, the first relation

$$h_{\underline{F}}^C = [\varsigma \, h_c + (1 - \varsigma) h_{H_2}] \, \omega_{LH} + h_{LOX} \, \omega_{LOX}$$

results.

By definition

$$(3.80) \qquad h_c := h_{LH} + \Delta h_c$$

the specific coolant enthalpy h_c consists of the enthalpy difference Δh_c and the 'fuel enthalpy' h_{LH} under feed conditions. Clearly, this property refers to the end of the cooling jacket at a flow state just before the coolant's mixture with the fuel main flow. Then one obtains from the equation above the final expression

$$(3.81) \qquad \boxed{h_{\underline{F}}^C = \frac{1}{1+O/F} \, [\varsigma \, \Delta h_c + h_{LH} - h_{LOX}] + h_{LOX}}$$

for the sought specific enthalpy of the fluid mixture in state $\underline{F}$. The cooling effect drops out with $\varsigma \to 0$, and the equation reduces to equation (1.28).

The coolant share ς varies between zero and one; typical data lie approximately at $\varsigma = 0.2$. For $\varsigma = 0$, $h_{\underline{F}}^C$ is converted to $h_{\underline{F}} = h_{\underline{F}}^m \psi_{\underline{F}}^{-1}$ using equation (1.28). With $\psi_{\underline{F}}$, the mean molar mass of the fuel-oxidizer mixture in state $\underline{F}$ is denoted, according to

$$\psi^{-1} = \psi_{O_2}^{-1} + [\psi_{H_2}^{-1} - \psi_{O_2}^{-1}] (1 + O/F)^{-1} \quad .$$

Using equation (3.81), the MM yields the reference solution, i.e., a solution for isentropic flows along the usual sequence of states. The resulting temperatures in these states can be used as initial data for establishing the free constants introduced by definition (3.67) of the cooling capacity $\hat{q}(\theta)$. For an assertible treatment of the cooling problem, however, it must be taken into account that the introduction of the function $\hat{q}(\theta)$ implies the entropy difference $\Delta s_{\underline{C} \to \underline{T} \to \underline{E}}$. Using the Carathéodory-Planck equation (3.66) together with definition (3.67), the expression

$$(3.82) \qquad ds = \frac{dT}{T} \left[-\frac{q_{\underline{C}}^m}{RT_{\underline{C}}} \frac{1 + c\theta}{\theta} R \psi_{\underline{C}}^{-1} \theta_1 \right]$$

follows for the corresponding increment establishing the proportionality between ds and dT.

From the above discussion it is clear that equation (3.82) offers only a characteristic example for inferences of modeling. It may provide detailed information needed to increase the understanding of regenerative cooling processes involved in reacting flows.

Relationship (3.82) now serves to modify the 'nozzle differential equation' derived for isentropic changes of state.

One begins with equation (3.29), in which the zero on the left side is substituted with equation (3.82). The repetition of the derivation presented in Section 3.3 is elementary, since expression (3.82) appears simply as an additive supplementary term. Without going into details, the final result, or '**differential equation for cooled nozzles**', can be given as:

$$(3.83) \qquad \boxed{\frac{dy}{dx} = (1 - \Gamma_p)^{-1} \left(c_{p,\varsigma}^m/R + \Gamma_T + a' \left[\frac{q_{\underline{C}}^m}{RT_{\underline{C}}} \right] (1 + c\theta)e^{c\theta} \right)} \quad .$$

Equation (3.83) differs from the 'nozzle differential equation' (3.56) only by the last term on the right side. This term must also be supplemented with equation (3.41) for the speed of sound: it seems practical, therefore, to combine this supplementary term with the property $c_{p,\xi}^m/R$ in the expression Γ_C. The abbreviations Γ_p and Γ_T are defined with equation (3.38), and the free constant a of the auxiliary θ_1 function is coupled with a' by the expression $a' = a \, R \, \psi_{\underline{C}}^{-1}$.

Equations (3.69) and (3.83) modify the MM in the case of regenerative cooling of the Laval nozzle. The choice of definition (3.67) for $\hat{q}(\theta)$ is made arbitrarily. In order to keep the discussions from getting too abstract and to offer an exemplary qualitative feature of a nozzle cooling modeling and its constructive elements, calculation of a concrete case has been preferred. The exponential form (3.67) is noticeable in equations (3.68), (3.77) and (3.83) because free constants (such as a and c) of the selected modeling function θ_* appear. These constants, in turn, can be coupled with the gas temperatures in states $\underline{C}$, $\underline{I}$ and $\underline{E}$ through simple modeling requirements. In contrast, equations (3.69) to (3.74) and (3.76) to (3.81), also important for the modeling, are independent from the mathematical form of the modeling function θ_*.

In general, it seems that the cooling problems of rocket engines can be treated theoretically with considerable flexibility, given an appropriate adaptation of the MM. The first exemplary calculations of this procedure are given in the final chapter.

Although the results presented here should be considered tentative, they have been substantiated by the most recent heat transfer investigations made by FIEBIG & MITRA (1988). Both authors studied the flow and temperature fields of a frozen H_2-O_2-mixture in a slender Laval nozzle, using the so-called 'Slender Channel Approximation' of the Navier-Stokes equations in cylindrical coordinates. Current data of the HM 60/1 was used as operational parameters.

The authors claim that the combined influence of friction and cooling on the thrust is small, given a Reynolds number of $20 \cdot 10^6$ for the nozzle throat cross section. The primary deviations from the 3.5% higher thrust calculated with the MM are "caused by the finite chemical reaction rates in the nozzle which were not taken into consideration by the slender channel calculation." This interpretation requires further detailed investigation.

"To what extent may our actions contradict our thoughts?"
                                                    -T.S.W. Salomon-

4. Test of the MM with Data from LH-LOX Rocket Engines

4.1 Problems of Assessment

In the final chapter of this study, the MM is evaluated with the tailored computer program 'CHOPER', and the results are compared with appropriate data from available literature. Since 'CHOPER' was created
solely for calculating the sequence of states of the ICP in a rocket
engine H_2-O_2-combustion, only results from the proper NASA codes as
well as performance and other test data from pertinent high-performance
engines can be considered for a comparison. Table 2 in Part I contains
the most important data available for current rocket engines.

Such a comparison and its subsequent assessment are faced with two significant problems from the very beginning:

(1) The amount of suitable test data is too limited to allow utiliza-
    tion of the MM's full spectrum for comprehensive parameter varia-
    tions.

(2) The NASA codes, still serving virtually as a world-wide standard,
    apparently stifle most creativity and rationality in design and ex-
    perimental work for rocket engines. Such work is customarily reduc-
    ed to routine completion of standardized calculations and measure-
    ments and is wholly inadequate for future projects!

A fair assessment of the MM primarily depends on the success in over-
coming the obvious psychological inhibitions of users accustomed to the
old NASA codes. It should be noted that acceptance of the MM requires a
bias-free reevaluation of some conventional key properties and their

conceptions. Such terms as the gasdynamical expansion flow or specific vacuum impulse are relevant examples.

Present design practice calls for elevating the specific vacuum impulse I_{sp} of an engine as high as possible. H.O. Ruppe, for example, writes in his standard work (1982, p.75) on the "modern rocket engine" ASE (see Table 2 in Part I): "It appears to me that the I_{sp} is a bit too high – 470s I believe. At any rate, that is the highest value (476s; D.S.) that has yet been quoted for an O_2/H_2 engine." This assessment is symptomatic: a difference of somewhat more than 1% amazingly deserves such a remark! The remark is absolutely correct, assuming that the actual target function of every given engine – the (vacuum) thrust S_E – is via definition

$$(4.1) \qquad\qquad S_E := \dot{m}_F \, I_{sp}$$

directly expressed by I_{sp} – because the steady mass flow rate $\dot{m}_F$ in the nozzle throat cross section is treated (within certain limits) as a free and flexible operating parameter.

This condition is found without exception in all pertinent NASA calculating modes as well as in the NASA-Lewis Code. It is also the basis of the new comparative investigation made by P. Kramer, in which the author makes an interesting attempt to list and weigh the relevant advantages and disadvantages of all current engine types, referring to their flight altitude and Mach number. Once again the measures for realizing and optimizing the I_{sp} receive major emphasis. In a remark by P. Kramer the goals of such studies – in one respect – become quite evident: "Thus it is the objective ... to reach the thrust density D_F and specific impulse I_s lying between those of rockets and air-breathing engines ..." (KRAMER 1987, p.62).

This study, on the other hand, offers relevant valid results with comprehensive practical and economical consequences, which strongly relativize the importance of I_{sp} . The following statements briefly summarize the results:

(1) If the geometric data, feed pressure and mass flow ratio of a rocket engine are established, the optimal thrust is clearly defined:
    the theory provides a mass flow rate $\dot{m}_{ev}$ assigned to the specific
    (vacuum) impulse I_{sp}.

(2) The specific (vacuum) impulse I_{sp} is primarily a material property
and is only marginally dependent on the type and mission of the engine.

The first statement notes that the mass flow rate can be neither freely
chosen as usual nor determined by the pressure in the feed system, as
long as one is looking for an optimal solution for the performance.[65]
In other words: I_{sp} is by no means - as normally interpreted in texts
and studies (see MIELKE 1986, p.473) - a "derived property" corresponding to $I_{sp} = S_E/\dot{m}_F$. This important point will be treated in greater detail in Section 4.3.

The second statement notes that I_{sp} primarily depends on the choice of
the chemical species which forms the fuel-oxidizer combination. At
technically relevant, comparable mass flow ratios, a H_2-O_2-mixture always has a higher specific impulse than, for example, a dimethyl hydra-
zine-tetroxide nitrate mixture (PIRUMOV & ROSLYAKOV 1986, pp.204-205).
This almost trivial fact implies a considerable influence of the mass
flow ratio O/F on the I_{sp}, since every change of the O/F ratio leads to
a new reactive mixture with different thermodynamic and kinetic data. A
functional coupling between I_{sp} and the geometric or mission-specified
parameters, however, is far less pronounced and not sufficiently clear;
dissipative effects often only minimally influence the specific vacuum
impulse (see PIRUMOV & ROSLYAKOV 1986, p.353).

The following Figure 4.1 is found in numerous recent publications. Influences of the flight altitude and Mach number on the I_{sp}-values are
practically irrelevant for the liquid-propellant and air-breathing
rocket and turbo-expander ramjet engines pertinent to this study.

Generally speaking, although I_{sp} is a characteristic parameter for a
rocket engine, it is by no means suitable for determining the engine's
optimal performance. S.S. Penner has previously emphasized the virtually negligible sensitivity of I_{sp} to modifications of relevant operative
conditions (see PENNER 1957, p.159).

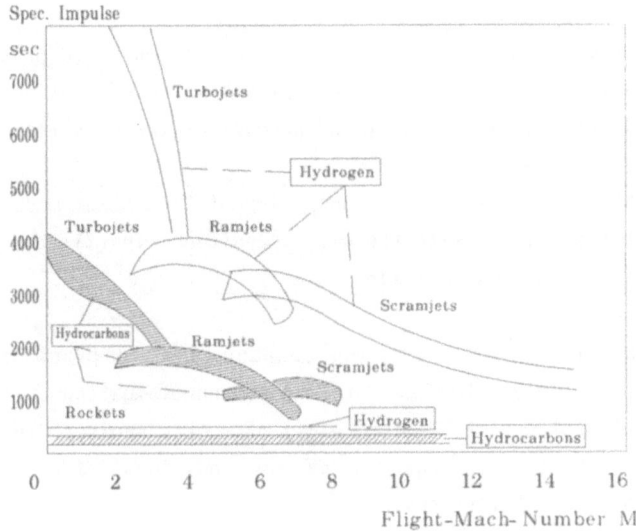

Figure 4.1: Specific Impulse of Combined Engines dependent on M

In the case of a H_2-O_2-combustion, the Munich Method essentially confirms the results of all professional calculating programs (NASA-Lewis Code, Brinkley, RAND algorithm, etc.) in which, with otherwise fixed parameters, the specific vacuum impulse I_{sp} is less dependent on the steady mass flow rate. R.J. Prozan, for example, bases his calculations of all prominent engine data on a mass flow rate of $\dot{m} = 1$ gm/s (PROZAN 1969, p.8). Insignificant differences naturally appear, but they generally lie under 1% . Thus we cannot confirm Prozan's negative evaluation (p.2) of the NASA-Lewis Code regarding erroneous (too high) I_{sp} values. In the following Table 4.1 , I_{sp} values (in seconds) for six rocket engines are compiled: they were calculated using the NASA-Lewis Code and the MM under the same conditions. All relevant engine data was previously compiled in Table 2 of Part I.

	J-2	J-2 S	SSME	ASE	HM7-B	HM60/1
NASA-Lewis Code	445.5	452.7	466.2	485.9	467.2	457.1
Munich Method	451.5	458.4	469.9	489	470.2	461

Table 4.1: Comparison of theoretically-obtained values of I_{sp}^{opt}

Summing up the results achieved so far, one notes that:

(i) given a specified fuel-oxidizer combination, feed pressure, mass flow- and expansion-ratio, the **optimal** value of the specific (vacuum) impulse is virtually independent of the mass flow rate.

(ii) for physical reasons, only one value of the steady mass flow rate can be compatible with the given conditions in (i): the eigenvalue $\dot{m}_{ev}$ as an **optimal mass flow rate**.

These two facts have surprising consequences for practical applications, particularly with regard to the common evaluation criteria (efficiency, thrust quality factor, etc.) of rocket engines. The following comparison scheme (with regard to vacuum conditions) offers a first impression of these inferences:

NASA-Methods		MM
given: $\left.\begin{array}{c} p_F \\ \dot{m}_F^c \end{array}\right\}$ * feed systems * $\left\{\begin{array}{c} p_F \\ \dot{m}_F^c \end{array}\right.$		
$S_E^{exp,c}$ * test * $S_E^{exp,c}$		
definition: I_{sp}^{eff} * $I_{sp}^{eff} := S_E^{exp,c}/\dot{m}_F^c$ * I_{sp}^{eff}		
theory: I_{sp}^{opt} $I_{sp}^{opt,c}$		
$!$ $\dot{m}_{ev}^c$		
definition: $S_E^{opt} := I_{sp}^{opt}(\dot{m}_F/\dot{m}_F^c)\,\dot{m}_F^c$ $S_E^{opt,c} := I_{sp}^{opt,c} \cdot \dot{m}_{ev}^c$		
$\dot{m}_F/\dot{m}_F^c \equiv 1$		
$S_E^{exp,c}/S_E^{opt} = I_{sp}^{eff}/I_{sp}^{opt} := \eta_I$ $S_E^{exp,c}/S_E^{opt,c} = [I_{sp}^{eff}/I_{sp}^{opt,c}]\cdot$		
$\cdot[\dot{m}_F^c/\dot{m}_{ev}^c]$		
(thrust quality factor) $\rightarrow \; \epsilon_S = \eta_I^c \dfrac{\dot{m}_F^c}{\dot{m}_{ev}^c}$		

Rocket engine performance data which can be precisely determined by experiment is listed immediately beneath the heading 'NASA Methods'. Subsequently the list is supplemented with appropriate definitions and the
specific vacuum impulse computed with the NASA-Lewis Code. It is decisive that the test data refers to a **cooled** engine: differences to properties calculated for the isentropic standard case are marked with the
superscript c.
An impulse efficiency η_I is introduced as one of several quality factors (see BARRèRE et al. 1960, p.102) for purposes of evaluation.

Since the NASA-Lewis Code does not contain a theory of cooling, it is
customary to compare the test data with standard case data. In the
first column it is apparent that in such a comparison, the steady mass
flow rate $\dot{m}_F^c$ of the cooled system is logically established as equal of
the common flow rate $\dot{m}_F$.

In contrast, the MM offers the possibility of modeling both the engine's cooling and the standard case. In addition, the mass flow eigenvalue $\dot{m}_{ev}$ is available as important supplementary information for cases
both with and without cooling.

Since a modeling of the cooling apparently has too many degrees of
freedom for a general discussion, this problem has been treated using
the theoretical connections of Section 3.7 for the actual example of
the HM60/1 characteristics. This evaluation has been carried out so
that the cooling coefficient ß for the chosen modeling is equal to one.
The ß value (obtained from equation 3.77) is, however, less than two
magnitudes for the operating parameters of the HM60/1. Therefore, for
this value ß = 1 the greatest share of the energy transferred from the
combustion gases is not absorbed by the coolant, but rather lost to the
environment. In this way the effectiveness of regenerative cooling may
be clearly indicated in view of the bulk properties of the combustion
gases.

The First Law in the form of equation (3.72) touches on the core of the
problem: the integral corresponds to the total amount of coolant $Q_{\underline{C \rightarrow T \rightarrow E}}$
transferred between the combustion chamber and the nozzle exit; it be-

Protocol 1

THEORETICAL ROCKET PERFORMANCE WITH MUNICH METHOD FOR THE ROCKET ENGINE HM60/1
===
AC/AT = 2.4600   AT =   547.00 cm*cm   O/F = 5.700 MASSFLOW-RATIO
HO(LOX,95.5 K) = -12455.90 J/mol   Delta Hk(120 K - 34 K) =   10477.96 kW
HO(LH ,34.5 K) =   -8308.60 J/mol   Cooling Rate Zeta   =   20.00 %
Factors : a = 27.29   b =   1.00   c = -6.64   BETA =   1.00
.   .   .
Qc-t =   -54276.9 kW   Qt-e = -322544.8 kW   Qk,c-e = -376821.7 kW
S-start = 18.02 J/K/g

DATA   FLUID   COMB.   THROAT   EXIT

A/AT	:		2.46	1.00	45.00
P (bar)	[ZETA=0., Delta S=0.]:	103.600	99.589	62.130	0.178
P (bar)	[COOLING, Delta S=0.]:	103.600	99.601	62.056	0.178
P (bar)	[COOLING] :	103.600	98.433	70.457	0.169
T (K)	[ZETA=0., Delta S=0.]:		3471.	3291.	1221.
T (K)	[COOLING, Delta S=0.]:		3476.	3296.	1226.
T (K)	[COOLING] :		3476.	3332.	1058.
H (J/g)	[ZETA=0., Delta S=0.]:	-949.	-1037.	-2152.	-10401.
H (J/g)	[COOLING, Delta S=0.]:	-912.	-1000.	-2117.	-10387.
H (J/g)	[COOLING] :	-912.	-1026.	-1940.	-10879.
W (g/mol)	[ZETA=0., Delta S=0.]:	9.88	13.01	13.15	13.51
W (g/mol)	[COOLING, Delta S=0.]:	9.88	13.01	13.15	13.51
W (g/mol)	[COOLING] :	9.88	13.00	13.13	13.51
S (J/K/g)	[ZETA=0., Delta S=0.]:		18.01	18.01	18.01
S (J/K/g)	[COOLING, Delta S=0.]:		18.02	18.02	18.02
S (J/K/g)	[COOLING] :		18.02	17.96	17.46
RHO (g/m**3)	[ZETA=0., Delta S=0.]:	339848.	4490.2	2985.8	23.7
RHO (g/m**3)	[COOLING, Delta S=0.]:	339848.	4482.6	2976.6	23.6
RHO (g/m**3)	[COOLING] :	339848.	4440.1	3340.3	25.9
MACH-NUMBER	[ZETA=0., Delta S=0.]:		0.251	1.000	4.469
MACH-NUMBER	[COOLING, Delta S=0.]:		0.251	1.000	4.467
MACH-NUMBER	[COOLING] :		0.286	1.000	4.906
V (m/s)	[ZETA=0., Delta S=0.]:	5.5381	419.	1551.	4348.
V (m/s)	[COOLING, Delta S=0.]:	5.5250	419.	1552.	4353.
V (m/s)	[COOLING] :	6.2345	477.	1560.	4468.
KAPPA/GAMMA	[ZETA=0., Delta S=0.]:		1.2535	1.1556	1.2592
KAPPA/GAMMA	[COOLING, Delta S=0.]:		1.2542	1.1551	1.2589
KAPPA/GAMMA	[COOLING] :		1.2544	1.1544	1.2732
X - H2	[ZETA=0., Delta S=0.]:	0.7373	0.2796	0.2780	0.2818
X - H2	[COOLING, Delta S=0.]:	0.7373	0.2797	0.2780	0.2818
X - H2	[COOLING] :	0.7373	0.2798	0.2782	0.2818
X - H2O	[ZETA=0., Delta S=0.]:		0.6512	0.6717	0.7182
X - H2O	[COOLING, Delta S=0.]:		0.6504	0.6709	0.7182
X - H2O	[COOLING] :		0.6500	0.6688	0.7182
THRUSTvak	[ZETA=0., Delta S=0.]:		0.00	732.58	1144.91
THRUSTvak	[COOLING, Delta S=0.]:		0.00	731.53	1143.67
THRUSTvak	[COOLING] :		0.00	830.30	1315.30
SP.IMP.vak(s)	[ZETA=0., Delta S=0.]:		0.00	294.95	460.96
SP.IMP.vak(s)	[COOLING, Delta S=0.]:		0.00	295.22	461.55
SP.IMP.vak(s)	[COOLING] :		0.00	296.95	470.41

---

m-dot (kg/s) [ZETA=0., Delta S=0.]: 253.26; >m-dot (O2) (kg/s) : 242.56
m-dot (kg/s) [COOLING, Delta S=0.]: 252.66; >m-dot (H2) (kg/s) : 42.55
m-dot (kg/s) [COOLING]   : 285.11;>>>m-dot (H2)c (kg/s) :   8.51

comes

$$(4.2) \qquad Q_{\underline{C} \rightarrow \underline{T} \rightarrow \underline{E}} = \frac{\zeta}{1 + O/F} \Delta h_c$$

and, in the case of the HM60/1 characteristics, is

$$Q_{\underline{C} \rightarrow \underline{T} \rightarrow \underline{E}} = \frac{0.2}{1 + 5.7} \, 1231.15 = 36.75 \text{ kJ/kg} \quad ,$$

insofar as the data given in the computer printout (**protocol 1**) is used. This low amount, which would be sufficient to balance the cooling process within the desired coolant temperature range without further transfer losses, contrasts with a specific energy of 1322 kJ/kg given off by the hot gases. This value equals the ratio of the gigantic cooling power $\overset{\circ}{Q}$ of 376,821.7 kW to the equally impressive combustion gas flow rate $\dot{m}_F^C$ of 285.1 kg/s.

Although this example seems rather unattractive for a modeling of the cooling, it offers two important insights:

(1) even a large amount of coolant scarcely has an **entropic** effect on the sequence of states in the combustion gas; with a more balanced modeling of the cooled ICP (i.e. at ß ≪ 1), the specific entropy in the combustion gas remains practically constant.

(2) the 'cooling parameters' ζ, O/F and Δh_c, on the contrary, influence (in constrast to the modeling equations for the cooling capacity q) the sequence of states energetically through equation (3.81): the cooling affects the values of I_{sp} and $\dot{m}_{ev}$.

This influence is by nature non-linear. With only a 3% entropy loss despite the extraordinary cooling power $\overset{\circ}{Q}$, however, it appears justifiable (subject to further investigations) to convert linearly the effects concerning I_{sp} and $\dot{m}_{ev}$ into the data of the 'complete regenerative cooling'. In both cases the initial value h_F^C changed by the cooling is assumed to be identical. The effects mentioned refer to the differences between the respective I_{sp}- and $\dot{m}_{ev}$-values of the cooled and uncooled engine; the latter is the standard case. One then obtains with

$$\overset{\circ}{Q}{}^{opt} := \dot{m}_F^C \cdot Q_{\underline{C} \rightarrow \underline{T} \rightarrow \underline{E}} = -10,478 \text{ kW} \quad ,$$

as the **optimized** cooling rate, the simple proportionalities

$$(4.3) \qquad \Delta \dot{m}_{ev}^{c,opt} := \frac{\overset{\circ}{Q}{}^{opt}}{\overset{\circ}{Q}} \Delta \dot{m}_{ev}^c \quad ; \qquad \Delta I_{sp}^{c,opt} := \frac{\overset{\circ}{Q}{}^{opt}}{\overset{\circ}{Q}} \Delta I_{sp}^c$$

by definition; in doing so the differences are always referred to the standard case, i.e. to $\dot{m}_{ev}$ and I_{sp} respectively. With the data from protocol 1, in which the results for the standard and (isentropic) reference case (see Section 3.7) as well as for the actual modeling ($\beta = 1$) of the cooling technique are given, the two thermodynamically optimized performance parameters (with $\zeta = 0.2$) become

$$\dot{m}_{ev}^{c,opt} = 254.1 \text{ kg/s} \quad ; \quad I_{sp}^{c,opt} = 461.2 \text{ s} \qquad 66$$

for the HM 60/1 characteristics: it is evident that these reference values differ only minimally from those of the standard case.

For the following evaluations, the last definition of the comparison scheme given above will thus be used in the form

$$(4.4) \qquad \boxed{\epsilon_S := \eta_I(\dot{m}_F^c / \dot{m}_{ev})} \quad ; \quad \eta_I := I_{sp}^{eff}/I_{sp}^{opt} \quad .$$

With this standard of quality – named the **thrust quality factor** – three different typical cases will be checked in the next section:

(α) $\quad \dot{m}_F^c/\dot{m}_{ev} \approx 1$ $\qquad\qquad$ optimal cooling: ASE

(β) $\quad \dot{m}_F^c/\dot{m}_{ev} \leq 1$ $\qquad\qquad$ optimal flow rate: HM 7

(γ) $\quad \dot{m}_F^c/\dot{m}_{ev} < 1$ $\qquad\qquad$ non-optimized flow rate: J-2S, SSME, HM 60/1, J-2 .

Naturally the distinguishing features listed above will at most indicate a preference. They are by no means sufficient for a comprehensive and reliable assessment of a rocket engine. To achieve such one also needs relatively banal information on the design requirements for the construction of an efficient and well-cooled Laval nozzle.

4.2 MM in Comparison with the Lewis Code and Test Data[67]

In order to present a fair comparison of the MM with the NASA-Lewis Code, one has to agree on corresponding inlet properties. This is particularly true for the specific entropy s_F of the fluid mixture. In the Lewis Code, this fluid property is established as equal to the specific entropy s_{ad} of the combustion gas obtained from the adiabatic flame temperature and the injection pressure p_F of the fluid mixture. It corresponds to the value obtained using the AFC method. For the MM, however, the value s_{ad} can at best (for reasons given in Section 3.6 and in Appendix 2) be considered as initial value for a realistic approximation of the specific fluid entropy s_F. For the MM, therefore, the assumption that $s_{ad} = s_F$ is a (bad) approximation and is used only in this chapter for comparative purposes.

In the following computer printout (**Protocol 2**), the ICP standard case is documented using the '**Advanced Space Engine**' (ASE) as an example. The experimental data necessary for the comparison was previously listed in Table 2 of Part I. The engine is primarily distinguished by two extreme geometric relationships: $\alpha := A_C/A_T = 3.66$ and $\epsilon := A_E/A_T = 400$. For clarity's sake, the most important characteristic values are compared below.

Design Parameter	Unit	Test Data	Lewis Code	MM
injection pressure	bar	140	140	140
chamber pressure	bar	137.9	–	137.8
mass flow rate	kg/s	19.18	19.18	19.38
vacuum thrust	kN	89.0	91.39	92.92
specific vacuum impulse	s	473.2	485.9	489
impulse efficiency η_I	–	–	0.974	0.968
mass flow rate-ratio	–	–	1	0.990
thrust quality factor ϵ_s	–	–	0.974	0.958

Table 4.2: Comparison of ASE data

The agreement between the values of the two theoretical methods is astonishingly good. The relatively high I_{sp}- values are unusual; even the effective vacuum impulse I_{sp}^{eff} from the experimental data is an exception. The fact that the experimental value for the mass flow rate is only minimally lower than the eigenvalue $\dot{m}_{ev}$ is an indication that the engine is cooled virtually without energy loss. The cooling technique causes an increase of pressure and velocity in the nozzle throat and thus (theoretically!) induces an increase of the steady mass flow rate (compare Protocol 1). The minimal deviations of both rates lead one to surmise that the manufacturer's engineers succeeded in operating the rocket engine at its optimal reference point. The three percentage points lacking for a perfect engine can be attributed to low dissipative effects. Their diverse causes can be identified with reasonable effort, examined individually and estimated.

A similar impressive example of excellent engineering talent is offered by the data of the 'Hydrogen Moteur 7-B' (HM7), with $\alpha = 2.88$ and $\epsilon = 82.9$. This data is compiled in **Protocol 3** according to results of the MM, and compared with test data and NASA calculations in Table 4.3:

Design Parameter	Unit	Test Data	Lewis Code	MM
injection pressure	bar	35.9	35.9	35.9
chamber pressure	bar	35	–	34.95
mass flow rate	kg/s	13.51	13.51	13.78
vacuum thrust	kN	59.4	61.9	63.57
specific vacuum impulse	s	448.3	467.2	470.2
impulse efficiency η_I	–	–	0.96	0.953
mass flow rate-ratio	–	–	1	0.980
thrust quality factor ϵ_s	–	–	0.96	0.934

Table 4.3: Comparison of HM7-B data

This rocket engine apparently still has some development potential. Although it approaches the theoretically optimal mass flow rate, fine adjustments, especially between dissipative processes within the combustion system and coolant flow, are nevertheless necessary.

Since the computer analysis of both the ASE and HM7 systems indicated only marginally deviating data, one may be tempted to believe that the NASA-Lewis Code is at least equivalent to the MM. Yet this impression is wholly deceptive, especially with regard to high-performance rocket engines! To emphasize this point, the respective results for two large

Protocol 2

THEORETICAL ROCKET PERFORMANCE WITH MUNICH METHOD FOR THE ROCKET ENGINE   ASE
===

AC/AT = 3.6600	AT = 31.97 cm*cm	S-START = 17.4063 J/K/g	T(LH) = 34.5 K	T(LOX) = 95.5 K
O/F = 6.000 MASSFLOW-RATIO	MASSFLOW RATE = 19.377 kg/s	C* = 2309.87 m/s	H(LH) = -4141.1 J/g	H(LOX) = -389.2 J/g
MULTIPLIER LAMBDA-E-BAR = -0.002166935			RHO(LH) = 68.0 kg/m**3	RHO(LOX) = 1138.0 kg/m**3

DATA	FLUID	COMB.	THROAT	EXIT	EXIT	EXIT	EXIT	EXIT	EXIT	EXIT	EXIT	EXIT	EXIT
A/AT	3.66	3.66	1.00	9.00	10.00	20.00	30.00	40.00	50.00	100.00	200.00	300.00	400.00
P (bar)	140.000	137.803	80.924	2.002	1.738	0.692	0.405	0.277	0.207	0.083	0.033	0.019	0.013
T (K)	------	3571.	3373.	1942.	1893.	1599.	1444.	1341.	1265.	1047.	856.	757.	692.
H (J/g)	-925.	-960.	-2105.	-8331.	-8491.	-9440.	-9920.	-10230.	-10456.	-11076.	-11590.	-11847.	-12011.
(H/HF)total	------	1.00	1.00	1.00	1.00	1.00	1.00	1.00	1.00	1.00	1.00	1.00	1.00
S (J/K/g)	17.41	17.40	17.26	17.26	17.26	17.26	17.26	17.26	17.26	17.26	17.26	17.26	17.26
W (g/mol)	10.24	13.53	13.68	14.11	14.11	14.11	14.11	14.11	14.11	14.11	14.11	14.11	14.11
RHO (g/m**3)	350380.	6282.0	3946.3	175.0	155.8	73.4	47.6	35.1	27.8	13.5	6.6	4.3	3.2
MACH NUMBER	------	0.159	1.000	3.263	3.337	3.831	4.129	4.347	4.520	5.091	5.728	6.139	6.451
V (m/s)	4.7263	264.	1536.	3848.	3890.	4127.	4241.	4314.	4366.	4506.	4618.	4674.	4709.
KAPPA/GAMMA	------	1.2606	1.1503	1.2157	1.2177	1.2315	1.2401	1.2467	1.2521	1.2705	1.2893	1.2996	1.3065
X - H2	0.7257	0.2482	0.2451	0.2438	0.2439	0.2440	0.2440	0.2440	0.2440	0.2440	0.2440	0.2440	0.2440
X - H2O	------	0.6756	0.6970	0.7557	0.7558	0.7560	0.7560	0.7560	0.7560	0.7560	0.7560	0.7560	0.7560
X - H	------	0.0299	0.0242	0.0004	0.0003	0.0000	0.0000	0.0000	0.0000	0.0000	0.0000	0.0000	0.0000
X - O	------	0.0028	0.0018	0.0000	0.0000	0.0000	0.0000	0.0000	0.0000	0.0000	0.0000	0.0000	0.0000
X - OH	------	0.0404	0.0300	0.0001	0.0001	0.0000	0.0000	0.0000	0.0000	0.0000	0.0000	0.0000	0.0000
X - O2	0.2743	0.0031	0.0020	0.0000	0.0000	0.0000	0.0000	0.0000	0.0000	0.0000	0.0000	0.0000	0.0000
THRUSTvac(kN)	0.00	0.00	55.63	80.33	80.93	84.39	86.07	87.14	87.90	89.96	91.61	92.41	92.92
THRUST (kN)	0.00	5.11	29.76	74.57	75.38	79.96	82.18	83.59	84.60	87.31	89.49	90.56	91.24
SP.IMP.vac(s)	0.00	0.00	292.75	422.74	425.90	444.07	452.93	458.56	462.58	473.40	482.06	486.29	488.96
SP.IMP. (s)	0.00	26.88	156.61	392.42	396.66	420.79	432.48	439.88	445.18	459.45	470.93	476.56	480.12

rocket engines,

* the **rocket engine J-2S** [68]
* the **'Hydrogen Moteur 60/1'**

are presented in the following tables. Once again, using Table 2 of
Part I, one obtains the necessary information in **Protocols 4 and 5.**

The first comparison concerns the J-2S engine, whose performance data
has been comprehensively documented (NN 1975):

Design Parameter	Unit	Test Data	Lewis Code	MM
injection pressure	bar	85.9	85.9	85.9
chamber pressure	bar	82.6	–	82.2
mass flow rate	kg/s	273.6	273.6	288.3
vacuum thrust	kN	1171	1215	1296
specific vacuum impulse	s	436.4	452.7	458.4
impulse efficiency η_I	–	–	0.964	0.952
mass flow rate-ratio	–	–	1	0.949
thrust quality factor ϵ_s	–	–	0.964	0.903

<u>Table 4.4</u>: Comparison of J-2S data

Currently there are even more variants for the HM 60/1; the following
comparison has been carried out:

Design Parameter	Unit	Test Data	Lewis Code	MM
injection pressure	bar	103.6	103.6	103.6
chamber pressure	bar	100	–	99.6
mass flow rate	kg/s	239.2	239.2	253.3
vacuum thrust	kN	1032	1073	1145
specific vacuum impulse	s	439.8	457.1	461
impulse efficiency η_I	–	–	0.961	0.954
mass flow rate-ratio	–	–	1	0.944
thrust quality factor ϵ_s	–	–	0.961	0.900

<u>Table 4.5</u>: Comparison of HM 60/1 data

Protocol 3

THEORETICAL ROCKET PERFORMANCE WITH MUNICH METHOD FOR THE ROCKET ENGINE HM7/B
===

AC/AT = 2.8800   AT = 88.50 cm*cm   S-START = 19.3348 J/K*g   T(LH) = 34.5 K   T(LOX) = 95.5 K
O/F = 5.300 MASSFLOW-RATIO   MASSFLOW RATE = 13.784 kg/s   H(LH) = -4141.1 J/g   H(LOX) = -389.2 J/g
MULTIPLIER LAMBDA-E-BAR = -0.00376735   C* = 2304.93 m/s   RHO(LH) = 68.0 kg/m**3   RHO(LOX) = 1138.0 kg/m**3

DATA	FLUID	COMB.	THROAT	EXIT	EXIT	EXIT	EXIT	EXIT	EXIT	EXIT	EXIT	EXIT
A/AT	2.88	2.88	1.00	9.00	10.00	20.00	30.00	40.00	50.00	60.00	70.00	80.00	82.90
P (bar)	35.900	34.951	21.011	0.508	0.440	0.173	0.100	0.068	0.051	0.040	0.032	0.027	0.026
T (K)	------	3301.	3123.	1732.	1686.	1404.	1256.	1159.	1087.	1030.	984.	946.	936.
H (J/g)	-985.	-1045.	-2195.	-8446.	-8606.	-9551.	-10023.	-10328.	-10547.	-10716.	-10853.	-10966.	-10995.
(H/HF)total	------	1.00	1.00	1.00	1.00	1.00	1.00	1.00	1.00	1.00	1.00	1.00	1.00
S (J/K/g)	------	19.33	19.32	19.20	19.20	19.20	19.20	19.20	19.20	19.20	19.20	19.20	19.20
W (g/mol)	9.52	12.24	12.37	12.70	12.70	12.70	12.70	12.70	12.70	12.70	12.70	12.70	12.70
RHO (g/m**3)	325360.	1558.8	1001.0	44.8	39.9	18.8	12.2	9.0	7.1	5.9	5.0	4.4	4.2
MACH NUMBER	------	0.207	1.000	3.266	3.343	3.859	4.174	4.405	4.590	4.745	4.880	5.000	5.033
V (m/s)	1.6622	347.	1556.	3863.	3904.	4139.	4252.	4323.	4373.	4412.	4443.	4468.	4475.
KAPPA/GAMMA	------	1.2570	1.1535	1.2333	1.2357	1.2517	1.2619	1.2696	1.2759	1.2813	1.2858	1.2894	1.2904
X - H2	0.7497	0.3217	0.3223	0.3321	0.3321	0.3322	0.3322	0.3322	0.3322	0.3322	0.3322	0.3322	0.3322
X - H2O	------	0.6091	0.6274	0.6677	0.6677	0.6678	0.6678	0.6678	0.6678	0.6678	0.6678	0.6678	0.6678
X - H	------	0.0363	0.0281	0.0002	0.0001	0.0000	0.0000	0.0000	0.0000	0.0000	0.0000	0.0000	0.0000
X - O	------	0.0020	0.0011	0.0000	0.0000	0.0000	0.0000	0.0000	0.0000	0.0000	0.0000	0.0000	0.0000
X - OH	------	0.0294	0.0202	0.0000	0.0000	0.0000	0.0000	0.0000	0.0000	0.0000	0.0000	0.0000	0.0000
X - O2	0.2503	0.0015	0.0008	0.0000	0.0000	0.0000	0.0000	0.0000	0.0000	0.0000	0.0000	0.0000	0.0000
THRUSTvac(kN)	0.00	0.00	40.04	57.29	57.71	60.11	61.27	62.00	62.52	62.92	63.24	63.50	63.56
THRUST (kN)	0.00	4.78	21.45	53.25	53.82	57.05	58.61	59.58	60.28	60.81	61.24	61.59	61.68
SP.IMP.vac(s)	0.00	0.00	296.21	423.83	426.93	444.69	453.26	458.67	462.51	465.44	467.79	469.72	470.22
SP.IMP. (s)	0.00	35.38	158.66	393.89	398.11	422.05	433.54	440.77	445.92	449.85	452.99	455.59	456.26

Both engines operate on the by-pass principle[69] and belong to the same performance class. It is immediately apparent that the J-2S engine by no means comes close to a theoretical optimum, because of the considerable difference between the actual mass flow rate and the eigenvalue. With the HM 60/1 as well, a 'reserve' of approximately 5% is available simply by optimizing this one factor. This is precisely the advantage of the MM over the performance-calculation of the NASA-Lewis Code! Evidently the method's basic practical advantage lies in its ability to indicate optimal operating characteristics for the feed system by precisely defining the relationship between injection pressure p_F and the mass flow rate.

This ability is relevant for selecting those parameters for which assumptions must be made, taking into consideration the technology available in the near future. Recently MANSKI & MARTIN (1988, p.2) have emphasized that "the turbo pump efficiency represents the gradient of technology improvements in the turbo-machinery."
The authors recommend a value of 75% for the 'isentropic efficiencies' of all turbines and pumps. In view of additional significant assumptions of total pressure losses from a turbo-machinery's exit to the main chamber, there are great challenges for creative engineers.

The results for the ASE and HM 7 offer impressive examples of the MM's advantage and indicate these performance characteristics can be calculated theoretically within rather narrow tolerances. In doing so, the strong interdependency between p_F and $\dot{m}_{ev}$ must be considered. The numerous operating parameters for the total configuration plus the feed system naturally require different design strategies. In the first place, high-pressure turbo pumps with performance levels approaching technical limits may be optimized. One can easily imagine the enormous research and development savings possible if these large turbo pumps could be designed from the very onset with the optimal operating characteristics $\hat{p}_F(\dot{m}_{ev})$.[70] One should be reminded that each O/F-combination represents, in principle, a completely new reactive fuel-oxidizer system. Thus the experimental establishment of the **feed characteristic** $\hat{p}_F(\dot{m}_{ev}|O/F)$ for high-performance engines requires huge investments - of time and money. Prior knowledge of such a relationship, therefore, has considerable advantages for mission profiles. This is illustrated by an example in the final section of this study.

Protocol 4

THEORETICAL ROCKET PERFORMANCE WITH MUNICH METHOD FOR THE ROCKET ENGINE J2-S
==

AC/AT = 2.3360   AT = 750.40 cm*cm   S-START = 17.9141 J/K/g   T(LH) = 34.5 K   T(LOX) = 95.5 K
O/F = 5.850 MASSFLOW-RATIO   MASSFLOW RATE = 288.340 kg/s   H(LH) = -4141.1 J/g   H(LOX) = -389.2 J/g
MULTIPLIER LAMBDA-E-BAR = -0.006178325   C* = 2235.53 m/s   RHO(LH) = 68.0 kg/m**3   RHO(LOX) = 1138.0 kg/m**3

DATA	FLUID	COMB.	THROAT	EXIT	EXIT	EXIT	EXIT	EXIT	EXIT	EXIT	EXIT	EXIT	EXIT
A/AT	2.34	2.34	1.00	9.00	10.00	12.50	15.00	17.50	20.00	25.00	30.00	35.00	39.80
P (bar)	85.900	82.240	51.411	1.271	1.103	0.818	0.642	0.523	0.438	0.326	0.256	0.209	0.176
T (K)		3474.	3307.	1903.	1855.	1756.	1679.	1616.	1562.	1476.	1408.	1353.	1308.
H (J/g)	-937.	-1032.	-2117.	-8338.	-8500.	-8826.	-9078.	-9281.	-9451.	-9721.	-9931.	-10100.	-10236.
(H/HF)total	-----	1.00	1.00	1.00	1.00	1.00	1.00	1.00	1.00	1.00	1.00	1.00	1.00
S (J/K/g)	17.91	17.89	17.75	17.75	17.75	17.75	17.75	17.75	17.75	17.75	17.75	17.75	17.75
W (g/mol)	10.09	13.22	13.38	13.81	13.81	13.81	13.81	13.81	13.81	13.81	13.81	13.81	13.81
RHO (g/m**3)	345149.	3765.6	2501.3	111.0	98.8	77.4	63.5	53.7	46.6	36.7	30.2	25.6	22.4
MACH NUMBER	-----	0.263	1.000	3.256	3.331	3.489	3.620	3.731	3.828	3.993	4.129	4.246	4.345
V (m/s)	4.7658	437.	1536.	3847.	3889.	3972.	4035.	4085.	4126.	4191.	4241.	4281.	4313.
KAPPA/GAMMA	1.2663	1.1482	1.2188	1.2210	1.2255	1.2292	1.2324	1.2353	1.2401	1.2442	1.2478	1.2509	
X - H2	0.7307	0.2640	0.2613	0.2627	0.2627	0.2628	0.2629	0.2629	0.2629	0.2629	0.2629	0.2629	0.2629
X - H2O	-----	0.6570	0.6797	0.7368	0.7369	0.7370	0.7371	0.7371	0.7371	0.7371	0.7371	0.7371	0.7371
X - H	-----	0.0332	0.0265	0.0004	0.0003	0.0001	0.0001	0.0000	0.0000	0.0000	0.0000	0.0000	0.0000
X - O	-----	0.0029	0.0018	0.0000	0.0000	0.0000	0.0000	0.0000	0.0000	0.0000	0.0000	0.0000	0.0000
X - OH	-----	0.0399	0.0289	0.0001	0.0001	0.0000	0.0000	0.0000	0.0000	0.0000	0.0000	0.0000	0.0000
X - O2	0.2693	0.0030	0.0018	0.0000	0.0000	0.0000	0.0000	0.0000	0.0000	0.0000	0.0000	0.0000	0.0000
THRUSTvac(kN)	0.00	0.00	828.74	1195.24	1204.20	1222.08	1235.70	1246.58	1255.56	1269.72	1280.54	1289.21	1296.12
THRUST (kN)	0.00	125.95	442.95	1109.38	1121.40	1145.32	1163.46	1177.92	1189.83	1208.58	1222.90	1234.35	1243.48
SP.IMP.vac(s)	0.00	0.00	293.07	422.68	425.85	432.17	436.99	440.84	444.01	449.02	452.85	455.92	458.36
SP.IMP. (s)	0.00	44.54	156.64	392.32	396.57	405.03	411.44	416.56	420.77	427.40	432.46	436.51	439.74

4.3 Variable Mixture Ratios and Reusability

The discussion of the cooling theory has made it clear that, under given circumstances, loss-free cooling of a rocket engine cannot be especially efficient, even if one were to use all the available liquid hydrogen ($\zeta \equiv 1$). This is due to the relatively high O/F-values, as one easily can see in equation (4.2). Since the cooling can only work if the highest coolant temperature does not exceed (or only minimally) the oxidizer's temperature at the entrance to the combustion chamber, an increase Δh_c of the liquid fuel's specific enthalpy is strictly limited. Theoretically, then, the only efficient possibility of noticeably lowering the temperature of the combustion gases lies in a significant reduction of the mass flow ratio O/F. Such a measure would promote a heat transfer at a low mean temperature level and thus prolong the strength and material stability of the combustion chamber and nozzle walls.

Many experts have previously pointed out the importance of this parameter as a variable. M. Valier, official inventor of the **rocket airplane** once again coming into vogue (OBERTH 1986, pp.277-278, and BRAUN & ORDWAY III 1979, p.206), was one of the first to accept the idea of H. Oberth regarding a H_2-O_2-combustion: he descriptively pointed out the effects of the "operating mixture" on temperature and gas velocity at the nozzle exit (VALIER 1928, pp.130 f). Recently BERGMANN (1984), MANSKI (1986) and KRAMER (1987) have preferred using this variable for a graphic presentation of relevant propulsion cycles influencing on the performance of rocket engines.

For cases in which the propellants themselves serve for pressurization, the mass flow ratio differs from the tank mixture ratio. This is because the fuel to oxidizer volume mixture ratio differs from its mass mixture ratio (see MANSKI & MARTIN 1988, p.2).

Investigations with the computer program 'CHOPER' have now revealed the surprising possibility of a new use for the mass flow ratio O/F. This variable can assure quantitative support for the practical applicability of the **'reusability'** concept in aerospace technology and obviously is of great importance for the future of the American Space Shuttle fleet (see FLETCHER 1988).

Protocol 5

THEORETICAL ROCKET PERFORMANCE WITH MUNICH METHOD FOR THE ROCKET ENGINE HM60/1
==

AC/AT = 2.4600   AT = 547.00 cm*cm   S-START = 18.0093 J/K/g   T(LH) = 34.5 K   T(LOX) = 95.5 K
O/F = 5.700 MASSFLOW-RATIO   MASSFLOW RATE = 253.262 kg/s   H(LH) = -4141.1 J/g   H(LOX) = -389.2 J/g
MULTIPLIER LAMBDA-E-BAR = -0.005321692   C* = 2237.57 m/s   RHO(LH) = 68.0 kg/m**3   RHO(LOX) = 1138.0 kg/m**3

DATA	FLUID	COMB.	THROAT	EXIT	EXIT	EXIT	EXIT	EXIT	EXIT	EXIT	EXIT	EXIT	EXIT
AJ/AT	2.46	2.46	1.00	9.00	10.00	12.50	15.00	20.00	25.00	30.00	40.00	45.00	50.00
P (bar)	103.600	99.589	62.130	1.519	1.318	0.977	0.765	0.522	0.388	0.305	0.208	0.178	0.155
T (K)	----.	3471.	3291.	1850.	1803.	1705.	1629.	1514.	1429.	1362.	1261.	1221.	1187.
H (J/g)	-949.	-1037.	-2152.	-8392.	-8552.	-8876.	-9127.	-9497.	-9766.	-9974.	-10281.	-10401.	-10504.
(H/HF)total	----.	1.00	1.00	1.00	1.00	1.00	1.00	1.00	1.00	1.00	1.00	1.00	1.00
S (J/K/g)	18.01	17.98	17.85	17.85	17.85	17.85	17.85	17.85	17.85	17.85	17.85	17.85	17.85
W (g/mol)	9.94	13.01	13.15	13.51	13.51	13.51	13.51	13.51	13.51	13.51	13.51	13.51	13.51
RHO (g/m**3)	339848.	4490.2	2985.8	133.3	118.7	93.0	76.3	56.0	44.1	36.3	26.8	23.7	21.2
MACH NUMBER	----.	0.251	1.000	3.269	3.344	3.504	3.636	3.847	4.013	4.152	4.375	4.469	4.553
V (m/s)	----.	419.	1551.	3858.	3899.	3982.	4044.	4135.	4199.	4248.	4320.	4348.	4371.
KAPPA/GAMMA	1.2535	1.1556	1.2231	1.2253	1.2299	1.2337	1.2399	1.2449	1.2492		1.2562	1.2592	1.2620
X - H2	0.7358	0.2796	0.2780	0.2817	0.2817	0.2817	0.2818	0.2818	0.2818	0.2818	0.2818	0.2818	0.2818
X - H2O	----.	0.6512	0.6717	0.7181	0.7181	0.7181	0.7182	0.7182	0.7182	0.7182	0.7182	0.7182	0.7182
X - H	----.	0.0306	0.0239	0.0002	0.0002	0.0001	0.0000	0.0000	0.0000	0.0000	0.0000	0.0000	0.0000
X - O	----.	0.0022	0.0012	0.0000	0.0000	0.0000	0.0000	0.0000	0.0000	0.0000	0.0000	0.0000	0.0000
X - OH	----.	0.0342	0.0240	0.0012	0.0001	0.0000	0.0000	0.0000	0.0000	0.0000	0.0000	0.0000	0.0000
X - O2	0.2642	0.0021	0.0012	0.0000	0.0000	0.0000	0.0000	0.0000	0.0000	0.0000	0.0000	0.0000	0.0000
THRUSTvac(kN)	0.00	0.00	732.58	1051.88	1059.65	1075.19	1087.01	1104.25	1116.54	1125.93	1139.65	1144.91	1149.44
THRUST (kN)	0.00	106.16	392.73	977.11	987.57	1008.41	1024.22	1047.18	1063.50	1075.95	1094.14	1101.11	1107.13
SP.IMP.vac(s)	0.00	0.00	294.95	423.51	426.64	432.89	437.65	444.59	449.54	453.32	458.84	460.96	462.79
SP.IMP. (s)	0.00	42.74	158.12	393.40	397.62	406.01	412.37	421.61	428.18	433.20	440.52	443.33	445.75

As previously mentioned, R.J. Prozan's salient investigations were not
only prompted by certain incompatibilities in the NASA performance calculations but, above all, by combustion temperatures too high for reusable rocket engines. Unfortunately Prozan's procedure has proven untenable, and the temperature drop (relative to the data provided by the
NASA-Lewis Code) predicted by him (see Figure 1.4) is simply not realistic under the given circumstances. According to his calculations, the
absolute temperature drop should lie in the magnitude of 500 to 800 K,
and thus be fully adequate for eliminating the acute overheating encountered in the Space Shuttle's main engines. The gas temperature in
the nozzle throat cross section would be reduced from 3400 K to around
2800 K. Obviously such a significant drop would decisively increase the
service life of the combustion chamber and the Laval nozzle. RUPPE
(1982, p.76) presented an interesting graph illustrating this problem:
wall temperatures over 800 K correspond to a maximum life time of 10^3
seconds; 'reusability' (defined as life times well over 10^4 seconds),
however, requires mean wall temperatures under 600 K!

This problem of reducing the combustion gas temperature is discussed in
detail below using the Space Shuttle Main Engine (SSME) as an example.
It is recommended that this investigation be supplemented with suitable
boundary layer studies. With Reynolds numbers in the magnitude of 10^7
(respective to the local state in the nozzle throat), the differences
in the various calculated results (due to laminar or turbulent wall
boundary layers) are especially interesting (see GEROPP 1987).

Initially a comparative analysis of the SSME's thermodynamics is necessary. The results of this MM investigation are compiled in the following computer printout (**Protocol 6**). Table 4.6 presents a data comparison such as those documented for other rocket engines in previous sections of this study – based on the data in Table 2 of Part I (and on
supplementary details given by RUPPE 1982, pp.73-74). In the subsequent
discussions, the geometric parameters $\alpha = 2.96$, $\epsilon = 77.5$, and the mass
flow ratio O/F are relevant.

THEORETICAL ROCKET PERFORMANCE WITH MUNICH METHOD FOR THE ROCKET ENGINE SSME

```
AC/AT = 2.9600    AT = 538.10 cm*cm    S-START = 17.1640 J/K/g    T(LH)  =   34.5 K      T(LOX)   =    95.5 K
O/F = 6.000 MASSFLOW-RATIO    MASSFLOW RATE = 497.005 kg/s    C* = 2250.90 m/s    H(LH)  = -4141.1 J/g      H(LOX)   =  -389.2 J/g
MULTIPLIER LAMBDA-E-BAR = -0.003401752                                     RHO(LH) =   68.0 kg/m**3   RHO(LOX) = 1138.0 kg/m**3
```

DATA		FLUID	COMB.	THROAT	EXIT	EXIT	EXIT	EXIT	EXIT	EXIT	EXIT	EXIT	EXIT	EXIT
A/AT	:	2.96	2.96	1.00	9.00	10.00	15.00	20.00	30.00	40.00	50.00	60.00	70.00	77.50
P (bar)	:	207.900	202.536	123.407	3.038	2.637	1.537	1.050	0.615	0.421	0.314	0.247	0.202	0.176
T (K)	:	-----	3605.	3408.	1936.	1888.	1711.	1594.	1439.	1336.	1260.	1200.	1151.	1119.
H (J/g)	:	-925.	-983.	-2118.	-8351.	-8511.	-9085.	-9456.	-9934.	-10244.	-10468.	-10642.	-10783.	-10873.
(H/HF)total	:	-----	1.00	1.00	1.00	1.00	1.00	1.00	1.00	1.00	1.00	1.00	1.00	1.00
S (J/K/g)	:	-----	17.16	17.14	17.00	17.00	17.00	17.00	17.00	17.00	17.00	17.00	17.00	17.00
W (g/mol)	:	10.24	13.59	13.73	14.11	14.11	14.11	14.11	14.11	14.11	14.11	14.11	14.11	14.11
RHO (g/m**3)	:	350380.	9181.1	5979.8	266.3	237.1	152.4	111.8	72.5	53.5	42.3	34.9	29.7	26.7
MACH NUMBER	:	-----	0.205	1.000	3.272	3.346	3.634	3.840	4.139	4.357	4.531	4.676	4.801	4.886
V (m/s)	:	8.9057	340.	1545.	3854.	3895.	4040.	4131.	4245.	4317.	4369.	4408.	4440.	4460.
KAPPA/GAMMA	:	-----	1.2503	1.1560	1.2162	1.2182	1.2260	1.2318	1.2405	1.2470	1.2524	1.2570	1.2611	1.2638
X - H2	:	0.7257	0.2480	0.2449	0.2438	0.2439	0.2440	0.2440	0.2440	0.2440	0.2440	0.2440	0.2440	0.2440
X - H2O	:	-----	0.6827	0.7036	0.7558	0.7558	0.7560	0.7560	0.7560	0.7560	0.7560	0.7560	0.7560	0.7560
X - H	:	-----	0.0268	0.0212	0.0003	0.0002	0.0001	0.0000	0.0000	0.0000	0.0000	0.0000	0.0000	0.0000
X - O	:	-----	0.0024	0.0014	0.0000	0.0000	0.0000	0.0000	0.0000	0.0000	0.0000	0.0000	0.0000	0.0000
X - OH	:	-----	0.0376	0.0272	0.0001	0.0001	0.0000	0.0000	0.0000	0.0000	0.0000	0.0000	0.0000	0.0000
X - O2	:	0.2743	0.0026	0.0016	0.0000	0.0000	0.0000	0.0000	0.0000	0.0000	0.0000	0.0000	0.0000	0.0000
THRUSTvac(kN):		0.00	0.00	1431.72	2062.49	2077.82	2131.84	2165.95	2208.94	2236.24	2255.77	2270.73	2282.73	2290.33
THRUST (kN):		0.00	168.92	767.67	1915.39	1935.90	2007.80	2052.95	2109.65	2145.60	2171.31	2191.02	2206.82	2216.84
SP.IMP.vac(s):		0.00	0.00	293.74	423.15	426.30	437.38	444.38	453.20	458.80	462.80	465.87	468.33	469.89
SP.IMP. (s):		0.00	34.66	157.50	392.97	397.18	411.93	421.19	432.83	440.20	445.48	449.52	452.76	454.82

Design Parameter	Unit	Test Data	Lewis Code	MM
injection pressure	bar	207.9	207.9	207.9
chamber pressure	bar	202.9	–	202.5
mass flow rate	kg/s	471.5	471.5	497
vacuum thrust	kN	2090	2155	2290
specific vacuum impulse	s	452	466	470
impulse efficiency η_I	–	–	0.970	0.962
mass flow rate-ratio	–	–	1	0.949
thrust quality factor ϵ_s	–	–	0.970	0.912

Table 4.6: Comparison of SSME data

The impulse efficiency η_I , as a scale of the influence of dissipative effects, etc. (see RUPPE 1982, p.38) within the combustor-nozzle flow, is valued as expected. The feed aggregate – characterized by the coefficient $\dot{m}_F/\dot{m}_{ev}$ – is not, however, thermodynamically optimized. The thrust quality factor ϵ_s is above 0.9, which apparently indicates a result previously matched or surpassed by rocket engines belonging to a low performance class. Yet although the SSME still has considerable development potential, it cannot be easily utilized if the previously-mentioned problem of an adequate drop in the gas temperatures is to be simultaneously solved.

The problem was approached by first using the MM to investigate the sequence of states under systematic variation of the mass flow ratio O/F and the injection pressure, with constant geometric parameters α, ϵ and A_T . Of crucial importance is the fact that the optimal thrust can be determined through the simultaneous calculation of the specific (vacuum) impulse I_{sp} and the mass flow eigenvalue $\dot{m}_{ev}$. Since the temperatures in the states of the sequence also must be calculated, one can correlate all relevant properties with one another. Figures 4.2 to 4.5 show the results of this comparison (note: the mass flow ratio O/F is always chosen as abscissa):

* Figure 4.2 confirms the known relationship between I_{sp} and O/F (see KRAMER 1987, pp.98-99, MANSKI 1986, p.57): between the low O/F-values and the stoichiometric mixture at O/F = 7.94, one observes a flat yet clearly identifiable maximum at $I_{sp} \approx 6$. Within certain limits, this value is, like the total curve, virtually independent of pressure. It is notable that the theory verifies the experimentally-obtained value of O/F = 6 for standard operation conditions of the SSME.

* Figure 4.3 impressively shows the curve of the obligatory mass flow eigenvalue $\dot{m}_{ev}$ to be a dependent operating parameter. Although the typical S-shaped influence of O/F on $\dot{m}_{ev}$ is quantitatively relatively low, its dependence on the pressure is distinctive. With a rise in pressure of about 30%, the increase in the mass flow rate is disproportionately higher. This feed characteristic is a decisive information for the economical development, testing and operation of high-enthalpy propulsion systems – especially rocket engines (see SUTTON 1986, pp.153 & 231).

* Figure 4.4 shows the curve of the gas temperature in state $\underline{I}$ of the nozzle throat cross section as a dependent of O/F. This function is a direct consequence of the feed characteristic $\dot{m}_{ev}(p_F; O/F)$. It is virtually independent of the feed pressure p_F . The temperature $T_{\underline{I}}$ (in contrast, for example, to the specific vacuum impulse) varies significantly within the range of technically-relevant O/F ratios. When the SSME operates at O/F = 6, the gas temperature is in the range of T > 3400 K, seriously jeopardizing the 'reusability' of the main engine.

* Figure 4.5 is undoubtedly the most interesting graph: every temperature drop considered necessary would scarcely be possible if it had to be achieved at the cost of a corresponding loss of available thrust. Such fears, however, apparently are groundless: a reduction of the value O/F = 6 to O/F = 4 , for example, theoretically results in a reduction of the specific (vacuum) impulse by around 1% and the (vacuum) thrust by less than 7%. The temperature drop in such a case is more than 700 K, or approximately 20% lower than the operating

gas temperature now known for the SSME! Even the theoretical 7% reduction of thrust still warrants a value which is considerably higher than the currently-measured 2090 KN. One observes the same qualitative behavior for deviating feed pressures.

The sequence of figures is by no means typical for the SSME: it is more
characteristic for the results of the MM. The respective $\hat{S}_E(O/F)$-curves
of various rocket engines differ qualitatively from one another, particularly with regard to their marked S-contours.

An unusual result is thus registered for the SSME: the gas temperature
can be lowered approximately 20% (!), for example, without having to
accept a proportional reduction of thrust. To accomplish this, the operation point has to be shifted from O/F = 6 to O/F $\approx$ 4, if necessary
through engine controls, as was done in the case of the SATURN 1B rocket and the J-2 rocket engine of the S-4B upper stage.

Before this can be done, however, one has to clarify whether ignition
of the fuel-oxidizer mixture is possible at the high combustion chamber
pressures and for the increased hydrogen excess (compared to the "ideal
value of O/F = 6"). Unfortunately there is a serious lack of reliable
empirical information on the pertinent high-pressure reaction kinetics.
Estimations of the suitable ignition times presented (with reservations) are assumed to give a positive answer. (WILHELMI 1987).

An increase in the structural weight of the entire rocket system must
currently be accepted in order to accomodate the increased load of hydrogen. An effective lowering of the combustion gas temperature under
otherwise unchanged conditions is possible only with a higher percentage of hydrogen in the O/F ratio. **Without this temperature reduction,
the concept of reusability is highly questionable.** Since such a concept
is virtually indispensable for the future of space programs (see FLET-
CHER 1988), the MM, in obvious contrast to the NASA-Lewis Code, offers
new and reliable perspectives for design, experimentation and practice!
With its use, everything from future combined cycle engines with highly
complex combustion processes to hypersonic combustion systems can be
rationally designed and tested. At present, only the MM can reliably
accomplish such work!

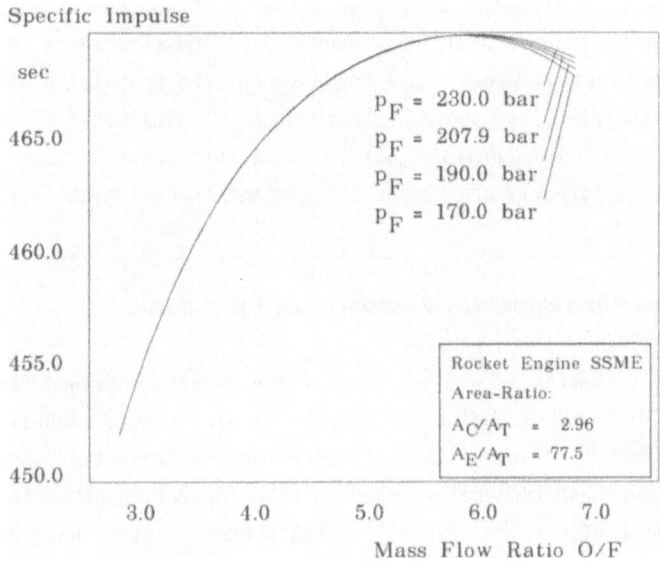

Figure 4.2: Specific Impulse dependent on O/F

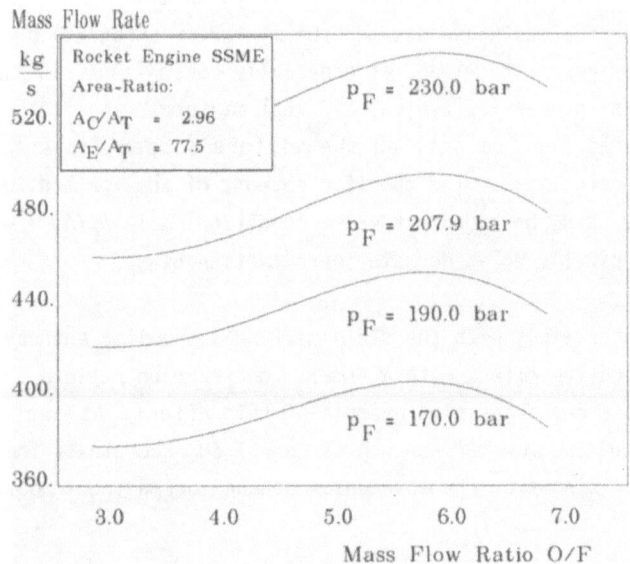

Figure 4.3: Steady Mass Flow Rate versus O/F

Such tasks are unusually complex, as MARGUET et al. (1970, p.4) noted as early as 1970 in an impressively clear statement: "It is shown, from projects under development at present, that at hypersonic speeds the interdependence between propulsion, aerodynamics and structures takes a primary importance and that, as a consequence, the propulsion studies cannot be dissociated from those of these other disciplines." HENDRICKS et al. (1986) recently pointed out that one must solve heat transfer problems which apparently are beyond the current technological limits.

4.4 Influence of the Operating Parameters on Flow States

Concluding this chapter is a discussion of the results obtained for the upper-stage J-2 rocket engine. In this particular study, emphasis was placed on an investigation of the influence thermal properties of state such as pressure and temperature have on other operating parameters. A distinctive lack of precise data on the substances involved has severely hindered an accurate analysis in the case of a LH-LOX **mixture.**

Although the individual thermal and caloric data of each component has been adequately measured for a large field of supercritical properties and described with reliable equations of state in the last ten years, apparently there is no known data on the mixture's extensive properties of state. This gap of knowledge is especially conspicuous in the case of the **excess** properties typical of real mixtures, and is the reason why the required precise data on the mixture's specific entropy s_F could not be calculated. **The specific entropy of the ignited gas mixture normally used, both under the AFC condition $\alpha := A_C/A_T \to \infty$ and in practice, is usually an inadequate approximation of s_F.**

This **first** uncertainty with the fluid mixture's specific entropy has an especially negative effect with a small **contraction ratio α** . The J-2 engine offers a characteristic example of this effect. As one can see from the results, low values such as $\alpha = 1.58$ accentuate the **second** uncertainty arising from the inaccuracy of the thermodynamic data used.

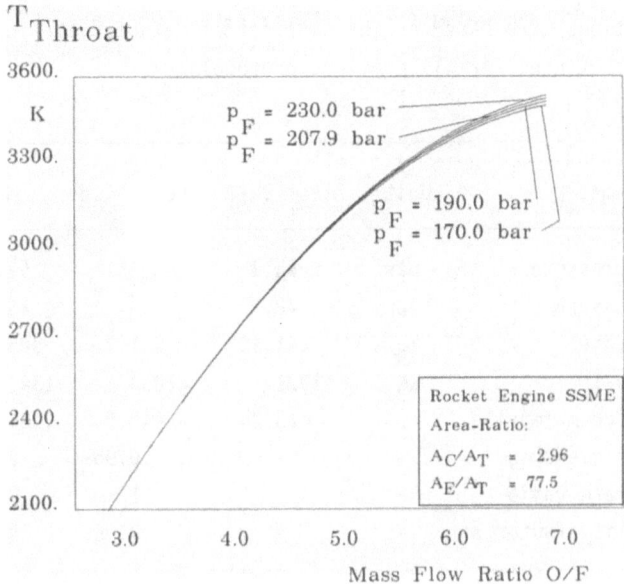

Figure 4.4: Gas Temperature in the Nozzle Throat vs O/F

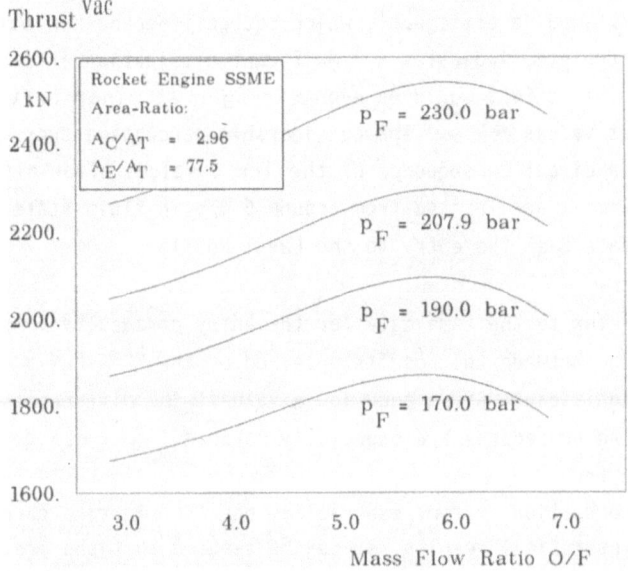

Figure 4.5: Vacuum-Thrust dependent on O/F

In order to facilitate a comparison of the J-2 data with those of other engines, Table 4.7 lists the results from tests, the Lewis Code and the MM. Calculations were based on the same set of data for the two components LH and LOX used for all previous protocols.

Design Parameter	Unit	Test Data	Lewis Code	MM
injection pressure	bar	53.1	53.1	53.1
chamber pressure	bar	48.7	–	45.3
mass flow rate	kg/s	241.2	241.2	303
vacuum thrust	kN	1001	1054	1341
specific vacuum impulse	s	423.2	445.5	451.5
impulse efficiency η_I	–	–	0.95	0.94
mass flow rate-ratio	–	–	1	0.80
thrust quality factor ϵ_s	–	–	0.95	0.75

Table 4.7: Comparison of J-2 data

At first glance the results are puzzling: apparently this best-known of all modern high-performance engines has by no means been perfected. The relatively low impulse efficiency, which scarcely depends on other operating constraints, indicates a significant dissipation effect in the flow system. This effect would be even stronger at minor contraction ratios than at values $\alpha \to \infty$. The considerable percentage pressure drop is naturally a direct consequence of the low α-value: after all, the flowing system is accelerated from around 5 m/s in fluid state to over 800 m/s in state $\underline{C}$ at the entry to the Laval nozzle!

The results refer to the same data for the entry parameters in state $\underline{F}$; the differences between the results obtained by the NASA-Lewis Code and the MM are significant. Given such low α-values, the differences in the two calculating procedures are especially notable. This can generally be said to be true.

In a H_2-O_2-combustion system, especially, one should take care interpreting the theoretical results. It can be assumed that the previously-mentioned condition $s_F \approx s_{ad}$ has an especially strong effect on the de-

termination of the mass flow eigenvalue in a J-2 engine. Since the actual (but unknown) s_F-value will always be smaller than the critical
value s_{ad} for $\alpha \to \infty$, one can also expect a lower value for $\dot{m}_{ev}$. Because
I_{sp} is scarcely affected by the choice of s_F , this lower value of $\dot{m}_{ev}$
could lead to a more favorable evaluation of the tested engine than is
given in Table 4.7. This situation is primarily a result of inadequate
information on the true value of s_F , yet is not even registered within
the scope of Lewis Code calculations if the data for the steady mass
flow rate is prescribed. On the contrary, the problem of lacking information arises every time, when a procedure involving the calculation of
$\dot{m}$ is used.

Protocol 7 presents a general survey of the J-2 engine data calculated
on the basis of the MM. Known thermodynamic properties of state for
both LH and LOX components have been chosen for the protocol headings.
In contrast to the reference case, this data is based on numerous NASA
analyses ('NASA case'). Although the temperature of LH in state $\underline{F}$ has
now been nearly reduced by half compared to the reference case (when
other properties of state of both components generally agree), one obtains data for the main operating parameters $\dot{m}_{ev}$, $I_{sp}(vac)$ and $S_E(vac)$
deviating only by fractions of a percentage point from the reference
case data. This insensitivity is obvious, since the difference in temperature causes only a minimal deviation of the caloric properties of
state from its standard value. At the same time, however, the excess
properties of the LH-LOX mixture influence **both** sets of data for state
$\underline{F}$ and implicitly the main operational parameters. The inherent influence of the excess entropy on the specific entropy of the multicomponent gaseous mixture, for example, is obviously quite considerable (see
Appendix 2).

The results compiled in Protocol 7 also demonstrate a striking defect
of the thermodynamic data used for the components of the combustion
gases. The mixture's specific entropy should remain unchanged through-
out the sequence of states. This holds true between states $\underline{F}$ and $\underline{C}$:
$s_F = s_{ad} = s_C$ is valid, whereby the relationship $s_F = s_C$ is used as a
condition for calculating the thermal properties of state in $\underline{C}$, and is
obtained with the MM calculation scheme (3.18). Such an isentropic condition is not explicitly employed for the respective changes of state

between $\underline{C}$ and $\underline{I}$ and $\underline{C}$ and $\underline{E}$. In other words: the physical premises of the MM (as for the NASA-Lewis Code) – $s_C = s_T$ and $s_T = s_E$ – are not used as part of the numerical calculating procedure for determining the properties of state in the above-named states. The nozzle differential equation is first introduced – derived from the isentropic condition ds = 0. Since the energy balance incorporated into the algorithm of the MM also assumes reversibility, the observed changes in the values of the absolute specific entropy cannot be explained as a procedural error of the MM. If one takes into consideration, however, that these s-values are 'subsequently' calculated and not a part of the iteration procedure, then the origin of the discrepancies in the s-values is obvious. It lies in the incompatibility of the polynomial coefficients used for calculating the specific enthalpies and entropies of all combustion gas components. These coefficients are filed in data banks of the NASA-Lewis Code and (for comparative reasons!) the computer program 'CHOPER'. Additional information is available in Section 3.6.

The more numerous the required incremental steps for solving the nozzle differential equation between two states are, the greater the resulting deviations. Based on the experience of assuming one such step for each temperature difference of approximately 5°C, the minimal differences in s between the states $\underline{C}$ and $\underline{I}$ indicate that relatively few steps (around 15 for the J-2 engine) are sufficient for a solution. This conclusion is valid for all rocket engines investigated and documented in the protocols.

The steadily increasing deviations between the state $\underline{I}$ and $\underline{E}$, however, indicate a considerably larger number of necessary incremental steps (approximately 450 for the J-2 engine!).

As the tests have shown, the discrepancies for rocket engines with low contraction ratios α increase at a greater rate than for those with higher α-values. The error is far less influenced by the area ratio $\epsilon := A_E/A_T$.

Although this incompatibility in the gaseous mixture's thermodynamic properties of state is not particularly significant for current applications of H_2-O_2-combustion in a high-performance engine, it should be clearly understood in order to detect other sources of error. This incompatibility should be avoided in the future even though it does not affect the theoretical background of the MM.

THEORETICAL ROCKET PERFORMANCE WITH MUNICH METHOD FOR THE ROCKET ENGINE   J-2

AC/AT = 1.5800	AT = 1099.00 cm*cm	S-START = 18.6562 J/K/g	T(LH) = 34.5 K		T(LOX) = 95.5 K
O/F = 5.552 MASSFLOW-RATIO		MASSFLOW RATE = 305.970 kg/s	H(LH) = -4141.1 J/g		H(LOX) = -389.2 J/g
MULTIPLIER LAMBDA-E-BAR = -0.022325881		C* = 1928.83 m/s	RHO(LH) = 68.0 kg/m**3		RHO(LOX) = 1138.0 kg/m**3

DATA	FLUID	COMB.	THROAT	EXIT	EXIT	EXIT	EXIT	EXIT	EXIT	EXIT	EXIT	EXIT	EXIT
A/AT	1.58	1.58	1.00	1.58	2.22	11.48	25.00	27.50	30.00	40.00	50.00	64.39	75.00
P (bar)	53.700	45.866	37.429	11.458	6.612	0.657	0.232	0.204	0.182	0.124	0.092	0.066	0.054
T (K)		3296.	3221.	2763.	2543.	1702.	1391.	1356.	1325.	1225.	1151.	1072.	1025.
H (J/g)	-962.	-1307.	-2164.	-4610.	-5617.	-8769.	-9780.	-9890.	-9988.	-10295.	-10517.	-10752.	-10886.
(H/HF)total		1.00	1.00	1.00	1.00	1.00	1.00	1.00	1.00	1.00	1.00	1.00	1.00
S (J/K/g)		18.66	18.53	18.47	18.44	18.40	18.40	18.40	18.40	18.40	18.40	18.40	18.40
W (g/mol)	9.79	12.67	12.85	13.08	13.14	13.21	13.21	13.21	13.21	13.21	13.21	13.21	13.21
RHO (g/m**3)	334548.	2121.4	1795.9	652.3	411.0	61.4	26.5	24.0	21.8	16.1	12.7	9.8	8.3
MACH NUMBER		0.503	1.000	1.877	2.204	3.440	4.016	4.089	4.156	4.382	4.563	4.774	4.905
V (m/s)	5.2670	831.	1550.	2701.	3051.	3952.	4200.	4226.	4249.	4320.	4371.	4425.	4455.
KAPPA/GAMMA		1.2634	1.1531	1.1793	1.1916	1.2318	1.2492	1.2515	1.2536	1.2609	1.2668	1.2739	1.2783
X - H2	0.7409	0.2961	0.2942	0.2959	0.2975	0.3004	0.3004	0.3004	0.3004	0.3004	0.3004	0.3004	0.3004
X - H2O		0.6277	0.6538	0.6848	0.6927	0.6995	0.6996	0.6996	0.6996	0.6996	0.6996	0.6996	0.6996
X - H		0.0351	0.0264	0.0116	0.0064	0.0001	0.0000	0.0000	0.0000	0.0000	0.0000	0.0000	0.0000
X - O		0.0025	0.0013	0.0002	0.0000	0.0000	0.0000	0.0000	0.0000	0.0000	0.0000	0.0000	0.0000
X - OH		0.0361	0.0232	0.0073	0.0033	0.0000	0.0000	0.0000	0.0000	0.0000	0.0000	0.0000	0.0000
X - O2	0.2591	0.0024	0.0011	0.0002	0.0000	0.0000	0.0000	0.0000	0.0000	0.0000	0.0000	0.0000	0.0000
THRUSTvac(kN):	0.00	0.00	885.68	1025.47	1094.92	1291.98	1348.75	1354.74	1360.04	1376.51	1388.26	1400.56	1407.49
THRUST (kN):	0.00	254.14	474.34	826.50	933.59	1209.05	1284.97	1292.94	1300.00	1321.91	1337.54	1353.92	1363.16
SP.IMP.vac(s):	0.00	0.00	295.16	341.75	364.89	430.57	449.49	451.48	453.25	458.74	462.65	466.75	469.06
SP.IMP. (s):	0.00	84.70	158.08	275.44	311.13	402.93	428.23	430.89	433.24	440.54	445.75	451.21	454.29

5. Summary Part II

Part II of this study is divided into two main sections. The first presents a comparison and critical commentary on current organisation- and calculation-procedures for equilibrium combustion in high-performance rocket propulsions, based on flow tube theory. The basic defects of today's virtually standard NASA-Lewis Code by S.Gordon & B.J.McBride are individually presented and analyzed. A comprehensive evaluation of the Code is given in Section 1.6.

The second main section is devoted to the principle theme of the study: the presentation and substantiation of the Munich Method (MM), as well as proof of its capabilities. At first a brief explanation is offered of an Alternative continuum Theory serving as the foundation of the MM. It describes the dynamics of real compressible fluid mixtures and offers for the first time a profound field theory for dissipative high-enthalpy flows. The resulting equations of motion differ basically from the (common) Navier-Stokes equation of motion. It is shown that the latter can at best be established physically for a pseudo mass-point model of an incompressible fluid. In comparison, the field equations of the Alternative Theory are consistently justifiable in all their details. Their first approximation offers a simple explanation for the amazing success of the Euler equation of motion in mapping viscous flow patterns. Traditional concepts of rational hydrodynamics are provably unable to provide such insights for establishing exacting standards.

The Alternative Theory is the physical basis of the Munich Method. It offers proof of how conditions for chemical equilibrium can be mathematically formulated for one-dimensional, steady, reversible flows of a multicomponent single-phase fluid mixture. This special case of constant mass flow density is elementary for the flow tube theory of reactive gas mixtures; it not only explains the pressure drop phenomenon,

but also describes it quantitatively as an exchange process between
energy forms of the Gibbs Fundamental Equation. It is notable that the
mixture's equilibrium composition at local pressures and temperatures
in the combustion chamber flow cannot be obtained through the corresponding equilibrium constants; they must be calculated from a non-
stoichiometric equilibrium formulation using Lagrange multipliers. It
is evident that the deviations between the results from this procedure
and those from the NASA-Lewis Code grow larger the smaller the ratio α
of the cross section areas of combustion chamber and nozzle throat.

Precisely determined properties of the state at the exit of the combustion chamber, together with the Mach number condition $M = 1$ in the narrowest nozzle cross section, allow the introduction of a so-called substitute process between the states $\underline{C}$ and $\underline{T}$. This virtual process is
definitely described by a non-linear normal differential equation independent of the actual contour of the Laval nozzle. An accurate representation of the speed of sound of a reactive gas mixture in chemical
equilibrium results from the derivation of this 'nozzle differential
equation'. The stoichiometric matrix needed for the exemplary case of a
hydrogen-oxygen combustion is discussed.

The 'nozzle differential equation' is also used to determine the properties of state in the nozzle exit cross section. Once they are known,
one can establish the isentropy of the reactive mixture's sequence of
states from the combustion chamber entry to the nozzle exit.

The Munich Method allows an equilibrium calculation for the process
realization along a combustion chamber-Laval nozzle facility as an
ideal-typical configuration. It also gives an optimal value of the mass
flow rate for a given feed pressure and known mass flow ratio of the
chosen fuel-oxidizer mixture. This result, especially useful for obtaining optimal operating characteristics of the engine's feed system,
indicates the practical superiority of the MM over the NASA-Lewis
Code's performance procedure. In this context it is remarkable that the
specific impulse of a propulsion system loses its usual priority as the
determining design quality!

Examinations of the performance data of well-known rocket propulsions with hydrogen-oxygen combustion also show two other results relevant to future combined cycle engines. First, one can systematically integrate the regenerative cooling of the engine into the overall design procedure. With a hydrogen-oxygen combustion, however, conventional cooling procedures cannot assure 'reusability' by protecting the excessively stressed combustion chamber and Laval nozzle construction units. A solution to this problem is indicated by this study: due to non-linear influences of chemical reactions on the energy conversions, engines can be operated with considerably lower mass flow ratios than usual without a significant loss of thrust. By doing so, one can simultaneously drop the combustion gas temperature by many hundreds of degrees and thus satisfy one of the essential requirements for 'reusability'!

This possibility has been investigated using the especially relevant example of the Space Shuttle-Orbiter's engine. A thermodynamical analysis of this rocket engine using the MM indicates a substantial development potential. The same is also true for the HM 60/1 used by the ARIANE 5. The best results were obtained for Rockwell International's 'Advanced Space Engine'(ASE): only the influence of dissipative effects were still registered. They are responsible for the remaining 4 to 5 percentage points and can be estimated using boundary layer equations as well as reaction kinetic relaxation formulas for chemical conversions in the divergent section of the Laval nozzle. It is notable that the flow rate chosen for the experimental determination of the ASE performance data deviates less than 1% from the mass flow eigenvalue obtained with the MM. Such reliable information cannot be obtained with the NASA-Lewis Code.

The J-2 rocket engine is apparently an exception due to its low contraction ratio. The MM analysis shows that the missing information for such factors as incompatibilities in thermodynamic data for the components in fluid or gas state can lead to substantial inaccuracies in performance prognoses. This lack of information is wholly unacceptable in a key technology such as H_2-O_2-combustion. As long as the information is not available, the J-2 engine should not be used as a reference model.

Table 5.1 summarizes the main results of this study as a comparison between all relevant qualities of both the NASA-Lewis Code and the MM.

Characteristic Informations	NASA-LC	MM
Thermodynamic Data		
Fluid State (LH-LOX)	unsolved	unsolved
Combustion (up to 200bar)	-	*
Equilibrium Program		
min $\hat{G}(T,p,n_k)$	+	*
Process Realization	not present	*
Configuration		
Ideal-typical	-	Minimum-Model
Numbers of Lengths	2	3
Expandable	no	yes
Performance Valuation		
Reversible Process	no	*
Chem. Equilibrium	-	*
Mixture Model	-	*
Throat-state	algebraic	NDE
Nozzle-Exit Pressure	heuristic	NDE
'Pressure-drop'	no	*
Steady Mass Flow Rate	indefinite	'eigenvalue'
Cooling Modeling	no	yes
Performance Simulation	precarious	developable

* problem adequate   + luxurious   - insufficient

Table 5.1: Comparison between Lewis Code and MM

The Munich Method, with its modifications, allows a comprehensive thermodynamic evaluation of rocket engine performance. Regenerative cooling can also be systematically incorporated into the calculations. With this innovative procedure, current and future designs of modern transonic to hypersonic combustion in combined cycle engines can be submitted to a rational and definitive theoretical analysis.

Bibliography

ADAMS, E.
On the Relevant Performance Parameters of an Ideal Rocket
Engine: Sensitivity Analysis; Internal Report (in German);
DFVLR - Bereich Raumfahrt - 01 TM 8603-AK/PA1; Köln 1987

BAEHR, H.D.
Thermodynamik -Eine Einführung in die Grundlagen und ihre
technischen Anwendungen- 2.umgearbeitete & erweiterte Auflage; Springer: Berlin 1966

BARRèRE, M.; JAUMOTTE, A.; FRAEIJS DE VEUBEKE, B.; VANDENKERCKHOVE,
J.Rocket Propulsion; Preface: M.Roy; Introduction: Th. v.
Kármán; Elsevier: Amsterdam 1960

BASEWICH, V.YA.
Chemical Kinetics in the Combustion Processes: A Detailed
Kinetics Mechanism and its Implementation; Prog. Energy Com-
bustion Sciences 13 (1987), pp.199-248

BAULCH, D.L.; DRYSDALE, D.D.; HORNE, D.G.; LLOYD, A.C.
Evaluated Kinetic Data for High Temperature Reactions; Vol.1;
Butterworths: London 1972

BAULCH, D.L.; DRYSDALE, D.D.; DUXBURY, J.; GRANT, S.
Evaluated Kinetic Data for High Temperature Reactions; Vol.3;
Butterworths: London 1976

BEJAN, A.
Entropy Generation Through Heat and Fluid Flow; A Wiley-
Interscience Publication; Wiley: New York 1982

BERGMANN, J.W.
Comparison of Ramjet Combustion Efficiencies (in German);
J. Flight Science and Space Research 8 (1984), pp.129-133

BIRKHOFF, G.
Hydrodynamics -A Study in Logic, Fact, and Similitude-
Princeton Univ. Press for Univ. of Cincinnati 1950

BRAUN, W.von; ORDWAY III, F.J.
The Rocket's Red Glare -An Illustrated History of Rocketry through the Ages- (in German); Pfriemer: München 1979

CALLEN, H.B.
Thermodynamics; Wiley: New York 1966

CLARKE, J.F.; McCHESNEY, M.
Dynamics of Relaxing Gases; Butterworths: London 1976

COOPER, L.P.; SCHEER, D.D.
Status of Advanced Propulsion for Space -Based Orbital
Transfer Vehicle-; Acta Astronautica 17 (1988), pp.515-529

DAVIS, Jr., J.G.; DIXON, S.C.
Beyond Simulation; Aerospace America July 1988, pp.38-42

DINGLER, H.
Max Planck und die Begründung der sogenannten modernen
theoretischen Physik; Ahnenerbe-Stiftung: Berlin 1939

DIXON-LEWIS, G.; WILLIAMS, D.J.
Comprehensive Chemical Kinetics; C.H.Bamford & C.F.K.
Tipper (Eds), Vol. 17; Elsevier: Amsterdam 1977

d'ESPAGNAT, B.
Grundprobleme der gegenwärtigen Physik; Wissenschaftstheorie,
Wissenschaft & Philosophie: S. Moser & S.J. Schmidt (Eds),
Vol.6; Vieweg: Braunschweig 1971

FALK, G.
Theoretische Physik –Auf der Grundlage einer allgemeinen
Dynamik–; Bd. I: Punktmechanik (1960), Bd. II: Allgemeine
Dynamik, Thermodynamik; Springer: Berlin 1966/68

FALK, G.; RUPPEL, W.
Mechanik-Relativität-Gravitation –Die Physik des
Naturwissenschaftlers–; Springer: Berlin 1973

FALK, G.; RUPPEL, W.
Energie und Entropie –Eine Einführung in die Thermodynamik–
Springer: Berlin 1976

FARMER, R.C.; PROZAN, R.J.; McGIMSEY, L.R.; RATLIFF, A.W.
Verification of a Mathematical Model which Represents Large
Liquid Rocket Engine Plumes; AIAA Paper 66 – 650, 1966

FIEBIG, M.; MITRA, N.K.
Model Computations of Laminar Dissipative Flows in a
Laval Nozzle; Internal Report (in German);
DFVLR –Bereich Raumfahrt– 01 TM 8603-AK/PA1; Köln 1988

FLETCHER, J.C.
Reusable Space Systems; J. Flight Sciences and Space
Research 12 (1988) pp.1-5

FRANCISCUS, L.C.; HEALY, J.A.
Computer Programs for Determining Effects of Chemical
Kinetics on Exhaust-Nozzle Performance; NASA TN D-4144;
NASA: Washington, D.C. 1967

FRANCK, E.U.
Fluids at High Pressures and Temperatures –Sixth Rossini Lecture 18.7.1986–; J.Chem. Thermodynamics 19 (1987), pp.225-242

GARDINER, W.C.,Jr.
Rates and Mechanisms of Chemical Reactions;
Benjamin: Menlo Park 1972

GARDINER, W.C.,Jr. (Ed.)
Chemistry of Combustion Reactions; Springer: New York 1983

GEROPP, D.
Boundary Layers in Supersonic Nozzles; Internal Report (in
German);DFVLR –Bereich Raumfahrt– 01 TM 8603-AK/PA1;Köln 1987

GLANSDORFF, P.; JAUMOTTE, A.; BALAND, J.
Sur la puissance disposible et le rendement de propulsion des
moteurs à réaction par jets. Bull. Acad. Roy. Belg., Cl.Sc.,
5e série, 41 (1955), pp.1264-1280

GOLDSTEIN, H.
Classical Mechanics; 8th Ed.(in German); Wissensch. Buchges.:
Darmstadt 1985

GORDON, S.; McBRIDE, B.J.
Computer Program for Calculation of Complex Chemical Equilib-
rium Compositions, Rocket Performance, Incident and Reflected
Shocks, and Chapman-Jouguet Detonations; NASA SP-273, Interim
Revision, March 1976; NASA Lewis Research Center: 1971

GRAUE, R.
Experimentelle Untersuchung chemischer und mechanischer Ab-
tragungsraten von Graphit in einem Hochenthalpiestrom; Fort-
schritts-Bericht VDI Ser.6:Nr.220;VDI-Verlag: Düsseldorf 1988

HANLEY, H.J.M.; EVANS, D.J.
A Thermodynamics for a System under Shear;
J.Chem.Phys. 76 (1982), pp.3226-3230

HEITMEIR, F.
Zur Theorie der Abbrandraten von Graphit in Hochenthal-
pieströmungen; VDI-Fortschr.-Bericht, Serie 6: Nr. 208;
VDI-Verlag: Düsseldorf 1987

HENDRICKS, R.C.; SIMONEAU, R.J.; DUNNING, J.W., JR.
Heat Transfer in Space Power and Propulsion Systems;
Mech. Eng. 108 (1986), pp.41-52

HÖGENAUER, E.
Space Transportation Systems (in German); J. Flight
Sciences and Space Research 11 (1987), pp.309-316

HOLUB, R.; VOŇKA, P.
The Chemical Equilibrium of Gaseous Systems;
Reidel Publication Comp.: Dordrecht 1976

JAMMER,M.
Concepts of Mass; Separate Edition (in German);
Wissensch. Buchgesellschaft: Darmstadt 1981

JISCHA, M.
Production Terms in Chemically Reaction Equilibrium Flows,
Discussed for a Binary Mixture Laminar Boundary Layer Flow.
Int. J. Heat Mass Transfer 16 (1973), pp.2261-2273

KAPPLER, G.
The New Generation of Aero Engines for Civil Applications (in
German); J.Flight Sci. and Space Research 12 (1988), pp.19-26

KELLER, J.U.
Semantic Comparison of some of the most Important Theories
of Irreversible Processes; Internal Report (in German);
DFVLR -Bereich Raumfahrt- 01 TM 8603-AK/PA1; Köln 1987

KRAMER, P.A.
Analyse des Antriebs- und Einsatzpotentials luftatmender Kombinationsantriebe für ballistische Raumtransporter; Dissertation Univ. Stuttgart 1987

KREUZER, H.J.
Nonequilibrium Thermodynamics and its Statistical Foundations -Monographs on the Physics and Chemistry of Materials-; Clarendon Press: Oxford 1983

LAVENDA, B.H.
Nonequilibrium Statistical Thermodynamics; A Wiley-Interscience Publication; Wiley: Chichester 1985

LIEPMANN, H.W.; ROSHKO, A.
Elements of Gasdynamics -Galcit Aeronautical Series-; Wiley: New York 1957

LIPPIG, V.
Computer Program 'CHOPER' for Calculation of Optimized Rocket Performance by H_2/O_2-Combustion; Internal Report (in German); DFVLR - Bereich Raumfahrt - 01 TM 8603-AK/PA1; Köln 1987

LUCAS, K.
Angewandte Statistische Thermodynamik; Springer: Berlin 1986

MANSKI, D.
Analyse und Optimierung kleiner Space-Shuttle-Antriebsplattformen; Dissertation TU Berlin 1984; DFVLR-Fb 84-28

MANSKI, D.
Effects of Engine Cycle Type on Payload Delivery of the Future European Launchers; J. Flight Sciences and Space Research 10 (1986), pp.51-61

MANSKI, D.; MARTIN, J.A.;
Evaluation of Innovative Rocket Engines for Single Stage Earth-to-Orbit Vehicles; AIAA/SAE/ASME/ASEE 24th Joint Propulsion Conference, July 11 - 13, 1988 / Boston, Mass.

MARGUET, R.; BARRÈRE, M.; CÉRÉSUELA, R.
Propulsion des Véhicules Hypersoniques; Note Technique No 169 ONERA: Chatillon 1970

MEBUS, H.G.
Berechnung von Raketentriebwerken; C.F. Wintersche Verlagshandlung: Füssen 1957

MEHRA, J. (Ed.)
The Physicist's Conception of Nature -Dedicated to P.A.M. Dirac on the occasion of his 70th birthday-; Reidel: Dordrecht 1973

MENNE, A. (Hg)
Philosophische Probleme von Arbeit und Technik; Wissensch. Buchges.: Darmstadt 1987

MIELKE, H.
    Lexikon der Raumfahrt und Weltraumforschung;
    transpress: Berlin 1986

MOORE, F.K. (ED.)
    Theory of Laminar Flows; Vol.IV: High Speed Aerodynamics and
    Jet Propulsion; Princeton Univ. Press: Princeton, N.J. 1964

NEBEL, R.
    Raketenflug; Raketenflugverlag: Berlin 1932

NEBENDAHL, P.; GRAUE, R.; KOLB, E.; TERHARDT, M.
    Thermodynamic Functions of LH, LOX, O_2,O,H_2,H,OH,H_2O,H_2O_2,HO_2
    (in German);DFVLR -Ber.Raumfahrt- 01 TM 8603-AK/PAI;Köln 1987

NEHRING, U.
    A Kinematic Approach to Unsteady Viscous Flows;
    14th ICAS-Congress Toulouse Sept. 1984, pp.494-503

NETTLETON, R.E.
    Shear-Rate Nonanalyticity in Simple Liquids;
    J. Non-Equilib. Thermodyn.12 (1987), pp.273-290

NOORDUNG, H.
    Das Problem der Befahrung des Weltraums – Der Raketenmotor;
    R.C. Schmidt: Berlin W62: 1929

N.N.
    JANAF Thermochemical Tables; First Addendum PB 168370-1
    Dow Chem. Midland, Mich, Proj.Dir.: D.R. Stull U.S. Dept.
    of Commerce 1966

N.N.
    JANAF Rocket Engine; Performance Methodology; Sample Cases;
    Chemical Propulsion Information Agency, John Hopkins Univ.:
    Silver Spring, Md. 1975

OBERTH, H.
    Wege zur Raumfahrt –Vorw. & Einf. zur Reprintausgabe E. Stuh-
    linger & H.Barth–; Klassiker der Technik; VDI-Verlag: Düssel-
    dorf 1986

OSWATITSCH, K.
    Grundlagen der Gasdynamik; Springer: Wien 1976

PENNER, S.S.
    Chemistry Problems in Jet Propulsion; Intern. Series of Mono-
    graphs on Aeronautical Sciences and Controlled Flight; Divi-
    sion III: Propulsion Systems including Fuels (Eds: E.R.Sharp,
    A.D. Baxter) Pergamon Press: London 1957

PETERS, N.
    Berechnung der Verbrennung in einer hypersonischen Grenz-
    schicht in der Nähe des chemischen Gleichgewichts; ZAMP 26
    (1975), pp.211-229

PETERS, N.; WARNATZ, J. (Eds)
Numerical Methods in Laminar Flame Propagation -A GAMM-work-
shop-; Notes on Numerical Fluid Mechanics Vol.6 (Ed.: K. För-
ster); Vieweg: Braunschweig 1982

PIRUMOV, U.G.; ROSLYAKOV, G.S.
Gas Flow in Nozzles; Springer Series in Chemical Physics,
Vol.29, Ed.: J.P. Toennies; Springer: Berlin 1986

PRAUSNITZ, J.M.
Molecular Thermodynamics of Fluid-Phase Equilibria; Inter-
nal Ser. in the Phys. & Chem. Sciences; Ed.: N.R.Amundson;
Prentice-Hall: Englewood Cliffs, N.J. 1969

PRIGOGINE, I; DEFAY, R.
Chemische Thermodynamik; VEB Dt. Verlag-Grundstoffindustrie:
Leipzig 1962

PROZAN, R.J.
Maximization of Entropy for Equilibrium Processes with
Emphasis on Rocket Engine Combustion Phenomena -Technical
Brief-; Lockheed Missiles & Space Comp., Huntsville, Ala.
HRECD 149342, Oct. 1969

PROZAN, R.J.
An Investigation of Equilibrium Concepts -Final Report-; Con-
tract NAS 8-34946 (CI-TR-0066) MSFC Huntsville, Ala.,Oct.1982

RICKEARD, D.
A One-Dimensional Flow Model for an Air-Augmented Rocket;
J. British Interplanetary Soc. 26 (1973), pp.18-27

ROWLINSON, J.S.
Liquids and Liquid Mixtures -Modern Aspects Series of Chemis-
try-; Gen. Ed.: F.C. Tompkins, Butterworths Scientific Publ.:
London 1959

RUPPE, H.O.
Die grenzenlose Dimension - Raumfahrt -; Band 2:
Werkzeuge und Welt, Econ: Düsseldorf 1982

RUPPE, H.O.
Basic Design Criteria for Space Payload Carriers (in German);
J. Flight Sciences and Space Research 9 (1985), pp.312-315

SÄNGER, E.
Raketenflugtechnik; Oldenbourg: München 1933

SCHÄFER, K.
Statistische Theorie der Materie; Vol. I,
Vandenhoeck & Ruprecht: Göttingen 1960

SCHILLING, W.; FRANCK, E.U.
Combustion and Diffusion Flames at High Pressures to 2000 bar
Ber. Bunsenges. Phys.Chem. 92 (1988) pp.631-636

SCHMIDT, E.
Thermodynamik; 10. revidierte Auflage; Springer: Berlin 1963

SERRIN, J. (Ed.)
New Perspectives in Thermodynamics; Springer: Berlin 1986

SHCHERBAKOV, S.A.
Thrust of a Contracting Nozzle; Translated from Izv. Akad.
Nauk SSSR, Mekh. Zhidk. Gaza, 6 (1983), pp.181-183; Plenum
Publ.Corp.(1984), pp.992-995

SMITH, W.R.; MISSEN, R.W.
Chemical Reaction; Equilibrium Analysis: Theory and
Algorithms; Wiley: New York 1982

STÖCKEL, K.
History of the Development of the Staged Combustion
Rocket Engine in Germany (in German);
J. Flight Sciences and Space Research 9 (1985), pp.1-14

STOLL, J.
Wärmeübergang und Filmkühlung in kompressibler Düsenströmung;
Dissertation Fak. Masch.-Wesen TU München 1987

STRAUB, D.
A History of the Glass Bead Game -Essay on the Role of Irre-
versibility in Physics-; 2nd Extended Version;Internal Report
A 02/86 St (in German);
Institute of Thermodynamics, Univ. Armed Forces Munich 1987

STRAUB, D.
On Microphysical Fundamentals of Thermofluiddynamics; Inter-
nal Report A 01/86 St (in German);
Institute of Thermodynamics, Univ. Armed Forces Munich 1987

STRAUB, D.; GRAUE, R.; HEITMEIR, F.; NEBENDAHL, P.; WURST, Th.K.
Fick's First Law Correction by an Exact Solution of the
Boltzmann Equation;
Propellants, Explosives, Pyrotechnics 12 (1987), pp.105-112

STRAUB, D.; LIPPIG, V.
Thermodynamical Aspects of Gasdynamical Relaxation Equations;
Seminary Series: Non-equilibrium Problems in Flow Mechanics
(Ed. K.Robert),pp.1-26 (in German);DFVLR: Porz-Wahn/Köln 1976

STRAUB, D.; WAIBEL, R.; LIPPIG, V.
On the Energy Theorem of Thermodynamics; Special Num-
ber 1977 on 70th Birthday of A. Walz (in German);
DLR-FB 77-16 (1977), pp.237-243

STREHLOW, R.A.
Combustion Fundamentals; McGraw-Hill Series in Energy,
Combustion and Environment; McGraw-Hill: New York 1984

SUTTON, G.P.
Rocket Propulsion Elements -An Introduction to the Engineering of Rockets-; Fifth Edition; A Wiley-Interscience Publication; Wiley: New York 1986

TRUESDELL, C.
Rational Thermodynamics with an Appendix by C.-C. Wang
Springer: New York 1984

ÜÇER, A.S.; STOW, P.; HIRSCH, Ch.; (Eds)
Thermodynamics and Fluid Mechanics of Turbomachinery
-Vol.II-; Nato ASI Series E: Applied Sciences - No.97A
Martinus Nijhoff: Dordrecht 1985

ULMER, K.; HÄFELE, W.; STEGMAIER, W.
Bedingungen der Zukunft -Ein naturwissenschaftlich-
philosophischer Dialog-; frommann-holzboog: Stuttg.1987

VALIER, M.
Raketenfahrt; 5. Auflage von 'Vorstoss in den Weltenraum -
eine technische Möglichkeit`; Oldenbourg: München 1928

VINCENTI, W.G.; KRUGER, C.H., Jr.
Introduction to Physical Gas Dynamics; Wiley: New York 1967

WEGENER, P.P.(ED.)
Nonequilibrium Flows: Part II; Gasdynamics:
A Series of Monographs; Dekker: New York 1970

WILHELMI, H.
Remarks on Chemism and Kinetics of H_2O_2-Combustion;
Internal Report (in German);
DFVLR -Bereich Raumfahrt- 01 TM 8603-AK/PA1; Köln 1987

WILLIAMS, F.A.
Combustion Theory -The Fundamental Theory of Chemically
Reacting Flow Systems-; Second Edition; Comb.Sci. & Engin.
Ser. (Ed.: F.A. Williams);
Benjamin/Cummings Publ.: Menlo Park 1985

WOODS, L.C.
The Thermodynamics of Fluid Systems;
Clarendon Press: Oxford 1975

WURST, Th.-K.
Transport Coefficients and Transfer Numbers for
H_2-O_2-Mixtures; Internal Report (in German);
DFVLR -Bereich Raumfahrt- 01 TM 8603-AK/PA1; Köln 1987

YEDLIN, A.K.
Calculation of the Injector End to Nozzles Stagnation Pres-
sure Ratio; Rocketdyne Division, North American Aviation,
PN 5114-2005, Canoga Park, Calif. 1965

ZIEREP, J.
Theoretische Gasdynamik; Dritte erw.u. überarbeitete Auflage;
Wissensch. + Technik: Taschenausgaben; Braun: Karlsruhe 1976

List of Relevant Symbols

* **Physical and Mathematical Properties**

A, A_r, a	area cross section, r-th reaction affinity, speed of sound
$\underline{A}$	formula matrix
b, $b_e^{\varnothing}$	squared mass flow density, initial mole number of the e-th element
$\mathbf{b}$	element quantity vector with entries b_e
c^*, c_p, c_v	characteristic velocity, specific heat at constant pressure and volume respectively
C_F	thrust coefficient
D, d	substantial time derivative, total differential operator
$\mathbf{D}$	deviator; traceless $\overset{\circ}{\mathbf{D}} := \nabla \mathbf{v}$
E; e	total energy, total number of elements; specific energy
e_{kin}, e_{pot}, e_{ϕ}	specific kinetic energy, potential energy, dissipation energy
F, f	field force, body force per unit mass
G, g_0	free enthalpy, standard gravitational acceleration
H, h, h_k	enthalpy, specific enthalpy, partial specific enthalpy of k-th component
I_{sp}, i	specific impulse (used in seconds and general unit)
$\mathbf{j}_e$, $\mathbf{j}_s$, $\mathbf{j}_k$	substantial current density of energy and entropy and diffusion of k-th species respectively
K, k_B	total number of species, Boltzmann constant
LH, LOX	liquid hydrogen, liquid oxygen
L	angular momentum
M, m_B, $\dot{m}$	Mach number, baryon mass, steady mass flow rate
N_k, n_k, $\dot{n}$	particle number variable, mole number of the k-th species, steady mole flow rate
$\mathbf{n}$	normal vector, component-quantity vector
O/F	oxidizer/fuel mass flow ratio

P, p	momentum, pressure
$Q, \dot{Q}$	heat, cooling power
$q, \dot{q}$	cooling capacity, heat current density
R, **R**	total number of reactions, material and universal gas constant respectively
r	position vector
S, s, s_k	entropy, specific entropy, partial specific entropy of the k-th species
S_E	engine's thrust
T, t, t_T	absolute temperature, time, empirical coefficient
V	volume
v, **v**	one-dimensional flow velocity, vectorial flow velocity
x_i^{can}	i-th canonical state variable
Y	generalized property of state
Z	compressibility factor
α	contraction ratio (A_C/A_T)
β, β_k	cooling coefficient, polynomial coefficient for thermodynamic functions of the k-th species
$\Gamma, \Gamma_T, \Gamma_p$	chemical production density, temperature- respective pressure- correction function
γ	modified isentropic exponent
δ	unit tensor
ϵ, ϵ_S	area ratio (= A_E/A_T); thrust quality factor
ζ	efficiency of combustion, coolant share
η, η_I	dynamical viscosity, impulse efficiency
θ, Θ	non-dimensional temperature, Lagrange multiplier, cooling function
κ	isentropic coefficient (heat capacity ratio)
$\lambda, \boldsymbol{\lambda}$	heat conductivity, Lagrange multiplier vector
μ, μ_k	shear viscosity, chemical potential of the k-th species
ν	kinematic viscosity

$\underset{\sim}{\nu}$	stoichiometric matrix with elements ν_{kr}
Ξ_T , Ξ_p	temperature correction, pressure correction
ξ_r	reaction coordinate of r-th reaction
$\overset{\bullet}{\pi}$	pressure tensor
ρ	(mass) density
σ	entropy production density
τ	stress tensor
ϕ	dissipation velocity
χ_k	mole fraction of k-th species
ψ, ψ_k , ψ_{max}	mean and component molar mass, respectively; efflux function
Ω_m, Ω_S	combustion chamber correction, thrust correction
ω_k	mass fraction of k-th species
∇	nabla operator
∂_t	partial time differential operator
$\mathscr{L}_s$	Lagrange function with respect to entropy

* **Superscripts**

c	due to cooling
m	molar property
exp	experimental property
inj	injection section
opt	optimum value
tot	total value
vac	due to vacuum
$[\rho]$	Legendre transformed variable with respect to ρ
^	denotes a function; $\hat{z}(x,y\vert a)$: z is a function of the variables x and y, and a functional of parameter a
$\emptyset$	standard value
~	total number of independent reactions

* **Subscripts**

e	e-th element: $e = 1(1)E$
k	k-th component or species: $k = 1(1)K$
S, s	due to thrust, isentropic change of state
U	ambient condition
$\underline{\omega}$, $\underline{F}$, $\underline{C}$, $\underline{T}$, $\underline{E}$	sequence of states indexed
ev	eigenvalue
*	local property of state in moving system

* **Abbreviations**

AFC	Adiabatic Flame Combustion
FAC	Finite Area Combustion
MM	Munich Method
PP	Prozan Procedure

* **Tensor Notation, Multiplications**

boldtype	vector and tensor, matrix with tilde
a b	dyad composed by vectors **a** und **b**
a·b	scalar formed by scalar product of vectors **a** and **b**
a × **b**	vector formed by vector product of vectors **a** and **b**
T·**b**	vector formed by scalar product of **b** and second order tensor T
$T_1 : T_2$	scalar formed by scalar product of second order tensors T_1 and T_2
∇s	vector formed by gradient of scalar s
∇·V	scalar formed by divergence of vector **V** or second order tensor **V**
∇v	tensor formed by gradient of vector **v**

In some cases brackets increase information: ( ) = scalar; [ ] = vector; { } = tensor.

APPENDIX 1: **Thermodynamic Analysis of the Simplified Model Gas**

The thermodynamic proof of a defined model gas using equation (3.1) is not a trivial matter if one wants to avoid misunderstandings in the gas dynamics procedure.

The **elementary ideal gases**, characterized by G. FALK (1968, pp.119 f), must first be examined. Each such gas (index i) has a specific heat (at constant pressure) of $c_{pi} := (5/2) R_i$ and is termed as polytropic or perfect. Its individuality is expressed by the material-specific gas constant R_i.

An elementary ideal gas i is defined by the Gibbs function

$$(3.a) \qquad \mu_i(T_i,p_i) := R_i T_i [\ln(p_i T_i^{-5/2}) - j_i] + e_{io} \qquad ;$$

the chemical constant j_i and the specific ground state energy e_{io} (at T = 0 K) appear in the expression. The constants are coupled with one another (FALK 1968, p.120): $e_{io} \approx \exp(2j_i/3)$.

The concept of a **mixture of elementary ideal gases** is fundamental. Such a 'simple mixture' is defined as follows: pressure p_i and the chemical potential μ_i of the i^{-th} elementary component are coupled with all other components through the three constraints

$$(3.b) \qquad p := \sum_i p_i \quad ; \quad T_i := T \quad ; \quad R_i := R \quad ; \quad (i = 1(1)I) \quad .$$

The pressure of the mixture and the partial pressure of the components are denoted respectively with p and p_i . Thermal equilibrium is established by the second constraint. Equation (3.a) – resolved with regard to p_i and introduced in equation (3.b) - yields

$$(3.c) \qquad p(T,\mu_1,\mu_2,\ldots) = \sum_i p_i(T,\mu_i) = T^{5/2} \sum_i a_i \exp\left[\frac{\mu_i - e_{io}}{RT}\right]$$

(with $j_i := \ln a_i$); the **free particle exchange** between the various components is expressed through the indexing of the chemical potential.

If this particle exchange is in a so-called diffusion equilibrium corresponding to $\mu_1 = \mu_2 = \ldots = \mu$, equation (3.c) is transformed into

$$(3.d) \qquad p(T,\mu) = a_1 T^{5/2} \exp\left[\frac{\mu - e_{10}}{RT}\right] \hat{Z}^{\dagger}(T) \qquad ,$$

where the abbreviation

$$Z^{\dagger} := \sum_i \left[\frac{e_{io}}{e_{10}}\right]^{3/2} e^{-(\epsilon_i/RT)} \qquad ; \qquad \epsilon_i := e_{io} - e_{10}$$

is introduced. The expression should be numbered so that the mark $i = 1$ stands for the elementary components with the smallest e_{io}-value. If this is done, then all energies ϵ_i have positive values. Equation (3.d) describes the thermodynamic behavior of a **simple** ideal gas in the sense of a simple phase. All real **pure substances** are composed of particle populations identified by different particle numbers but identical molar masses. Each such population is an individual component and relates to the same constant value ϵ; the index i thus indicates the characteristic distribution function $\hat{\epsilon}_i(N_i)$.

The sum $Z^{\dagger}$ is simply the sum of every component's particle N_i number divided by N_1 - as one immediately recognizes from the thermal equation of state for the elementary ideal gas

$$(3.e) \qquad p_i V := \left[\frac{N_i}{N_L}\right] M RT \qquad ;$$

from this it immediately follows $p_i/p_1 = N_i/N_1$; M denotes the molar mass for all components and N_L is the Loschmidt number (see FALK 1968, p. 122). By comparing $p = \sum p_i = p_1 \sum N_i/N_1$ with equation (3.d), one obtains

$$(3.f) \qquad \frac{N_i}{N_1} = \left[1 + \frac{\epsilon_i}{e_{10}}\right]^{3/2} \exp(-\epsilon_i/RT)$$

as a particle ratio for the i^{-th} and first components. In a diffusion equilibrium, these components are dependent only on the temperature. A generalization for N_i/N_{i+1} is simple.

Based on experience, the relationships for elementary ideal gases are only valid as normal limiting laws for real fluids if the constraint $RT \ll e_{10}$ is fulfilled. Since the ground state energy of the gas parti- cles is equal to the rest energy of its molecules, this constraint can usually be satisfied. As a practical consequence, the 'degeneracy' [*] $\gamma_i = (e_{i0}/e_{10})^{3/2}$ – which gives the number of the various elementary components with the same ground state energy e_{i0} – can be used as a good approximation of being one: $\epsilon_i/e_{10} \ll 1$ is thus valid.

The sum Z^t is then reduced to

(3.g)
$$\hat{Z}(T) = \sum_i \exp(-\epsilon_i/RT) \quad ;$$

this expression denotes the **internal partition function** of the simple mixture.

The simple mixture's properties of state which are of particular inter- est here – the specific entropy s and the specific enthalpy h – can be calculated using the mixture's Gibbs function $\hat{p}(T,\mu)$ and the equations (FALK 1968, p.123)

(3.h1)
$$\rho s = \left[\frac{\partial p}{\partial T}\right]_\mu$$

(3.i1)
$$h = \mu + Ts \quad ;$$

the mass density ρ is given by the thermal equation of state

(3.j)
$$p = RT\rho = (R/M)T\rho$$

for an ideal gas.

---

[*] the degeneracy γ_i equals the statistical weight to be assigned to each level of energy as a inference of quantum mechanics. The part- icles are indistinguishable from one another , and individual quan- tum states are equally likely. Simply expressed: γ_i quantifies the possibility that any macro-state may be realized by several micro- arrangements.

Using equation (3.d), one thus obtains the expression

$$(3.h2) \quad s = RT\left[\frac{\partial \ln p}{\partial T}\right]_\mu = \frac{5}{2} R - \frac{\mu - e_{10}}{T} + RT \frac{d}{dT} \ln \hat{Z}(T)$$

for the specific entropy.
After multiplication with T using equation (3.i), the expression immediately provides the sought relationship to the specific enthalpy

$$(3.i2) \quad h = \frac{5}{2} RT + RT^2 \frac{d}{dT} \ln \hat{Z}(T) + e_{10}$$

of the simple mixture.

A physical analysis of an ideal gas as a mixture of elementary components follows, using an especially simple example for illustration. Two principles will be examined here: L. Boltzmann's concept of thermodynamics (which considers **every entropy to be a mixture entropy**) and the fundamental thermodynamic concept of conversion processes. According to these concepts, the elementary components consist of those particles of a unified ideal gas (with R and M) "which are found in the **same internal molecular state.** Index i, which identifies the various elementary components, also marks the internal molecular state. The ground state energy e_{io} is the internal energy of a molecule which rests in this internal state. The ϵ_i, then, is the excitation energy of the i^{-th} internal molecular state...." (FALK 1968, p.124).

In a monatomic elementary ideal gas all particles (atoms) are generally in the ground state, since the excitation energies of an atom are usually too high, compared to RT, at technically-operable temperatures.

On the other hand, an elementary ideal gas need not be monatomic: it suffices if all molecules of the gas are in the same internal state.

In order to illustrate the numerous possibilities of such states, one can examine a (r + 1)-component mixture in which one component has energy e_{10} and r-components have the same ground energy e_{20}.
Using the expression $\epsilon := e_{20} - e_{10} \geq 0$, one obtains the internal partition function expressed as

$$(3.k) \quad \hat{Z}(T) = 1 + r\, e^{-\epsilon/RT} \quad .$$

Together with the solution $\hat{\mu}(T,p)$ (resolved with equation (3.d))

$$(3.1) \qquad \mu(T,p) = RT[\ln(pT^{-5/2}) - j] + e_{10} - RT \ln \hat{Z}(T) \quad,$$

and considering equation (3.i2) and (3.h2), one arrives at the temperature-pressure functions for the specific enthalpy h

$$h = \frac{5}{2} RT + \epsilon\left[1 + \frac{1}{r} e^{\epsilon/RT}\right]^{-1} + e_{10} \quad,$$

and

$$s = \left[\left(\frac{5}{2} + j\right) - \ln\left(pT^{-5/2}\right) + \ln\left(1 + r\, e^{-\epsilon/RT}\right)\right] R + \frac{\epsilon}{T}\left(1 + \frac{1}{r} e^{\epsilon/RT}\right)^{-1}$$

for the specific entropy s.

One can easily show (see FALK 1968/2, pp.87 f) that in cases when diatomic molecules have **fully excited rotational degrees of freedom**, both equations have to be slightly modified. The relationships

$$(3.m1) \qquad \boxed{h = \frac{\kappa}{\kappa-1} RT + R\theta\left[1 + \frac{1}{r} e^{\theta/T}\right]^{-1} + e_{10}}$$

$$(3.n1) \qquad \frac{s}{R} = \frac{\kappa}{\kappa-1} \ln T - \ln p + \frac{\hat{s}^{*}(T)}{R} + \left(\frac{7}{2} + j - \ln \theta\right)$$

now contain only the influences of the electronic and vibrational excitations on the temperature dependency of h and s, if the common numerical value $\kappa = 7/5 = 1.4$ is valid for the isentropic coefficient $\kappa := c_p/(c_p - R)$.

Here the dimensionless temperature function $\hat{s}^{*}(T)/R$

$$(3.n2) \qquad \frac{\hat{s}^{*}(T)}{R} := \ln\left[(1 + r\, e^{-\theta/T})\, \exp \frac{\theta}{T}(1 + \frac{1}{r} e^{\theta/T})^{-1}\right]$$

generally corresponds to the mixing entropy of the model gas. Using $\theta := \epsilon/R$ as an abbreviation, this non-dimensional excitation energy (in the unit K) indicates, above all, vibrational relaxation. The following relationship in closed algebraic form

(3.o)

$$\frac{P_{\zeta+1}}{P_\zeta} = \left(\frac{T_{\zeta+1}}{T_\zeta}\right)^{\kappa/\kappa-1} \frac{\left[1 + r\,e^{-\theta/T_{\zeta+1}}\right]\exp\left[\frac{\theta}{T_{\zeta+1}}\left[1 + \frac{1}{r}\,e^{\theta/T_{\zeta+1}}\right]^{-1}\right]}{\left[(1 + r\,e^{-\theta/T_\zeta}\right]\exp\left[\frac{\theta}{T_\zeta}\left[1 + \frac{1}{r}\,e^{\theta/T_\zeta}\right]^{-1}\right]}$$

for isentropic changes between the states ζ and $\zeta+1$ clearly shows that the incompletely excited excitation energies cause deviations from the classic 'adiabatic law' $pT^{(\kappa-1)/\kappa}$ = constant. Equation (3.o) may be denoted as **Extended Laplace-Poisson Law**.

The example given here was chosen to illustrate an important aspect: relationship (3.o) doesn't describe a state (such as the thermal equation of state), but rather a process between two states. A conversion of energy occurs in this process and can be described, for example, with parameter r. A well-known and important example of how r remains unchanged is offered by normal hydrogen, a mixture of 'para' and 'ortho' hydrogen (see FALK 1968/II, pp.90 f).

In normal conditions neither fluid component interchanges. Under catalytic influence, however, a free particle interchange can be realized, and normal hydrogen arises.

The model gas sought here for an exemplary gas dynamic calculation should show the simplest possible thermodynamic behaviour between state F at the inlet of the combustion chamber and state E at the outlet of the Laval nozzle, and yet exhibit the characteristic behaviour of equilibrium processes.

From equation (3.ml), combined with the definition

(3.p) $\qquad\qquad r\,e^{-\theta/T} := \gamma$

containing the limiting cases $\gamma \to r$ for $T \to \infty$ and $\gamma \to 0$ for $T \to 0$ K respectively (and unbounded r), one first obtains

(3.m2)
$$h = \frac{\kappa}{\kappa-1} RT + \frac{1}{1+\gamma^{-1}} R\theta + e_{10}$$

for the specific enthalpy of the simplified model gas corresponding to equation (3.1) in Part I. It is hypothetically assumed that the relaxation parameter γ, representing the changes in r during a process between two states, is influenced by both the local values of the variables of state and the **history** of hidden affinities. Parameters like γ that relating a process to the history of an affinity often may be reduced to so-called response functions (see WOODS 1975, p.268).

Once the enthalpy difference between states ς and $\varsigma+1$ is determined, relationship

(3.q)
$$\Delta h = h_{\varsigma+1} - h_\varsigma = \frac{\kappa}{\kappa-1} \left[T_{\varsigma+1} - T_\varsigma \right] R +$$

$$+ \left[\left[1 + \gamma_{\varsigma+1}^{-1} \right]^{-1} - \left[1 + \gamma_\varsigma^{-1} \right]^{-1} \right] R\theta$$

follows, assuming that the parameter-value γ in state $\varsigma+1$ differs from the value in state ς; equation (3.m1) already points to this hypothesis concerning r or γ respectively with regard to real processes. This is especially evident when dealing with irreversible mechanisms.

It is a major simplification for the **model gas** (Index ^) if one can presume that γ depends neither on T nor on p, but only on the **history**. In this case, the expression

(3.r)
$$\lim_{\Delta s \to 0} \gamma_{\varsigma+1} = \gamma_\varsigma \equiv \hat{\gamma} \,,$$

is valid for the **isentropic** limit, and there are two major inferences:

(1) The second term on the right side of equation (3.q) is dropped: the difference in enthalpy depends (regardless of the difference) only on a constant κ value as a parameter. This result is significant

for the gas dynamic energy equation, since the latter now no longer requires explicit knowledge of γ.

(2) The **Extended Laplace-Poisson Law** can now be expressed

(3.s)
$$\frac{p_{\varsigma+1}}{p_\varsigma} = \left[\frac{T_{\varsigma+1}}{T_\varsigma}\right]^{\kappa/(\kappa-1)} \exp\left[\frac{\theta}{1-[\hat{\gamma}]^{-1}}\left[\frac{1}{T_{\varsigma+1}} - \frac{1}{T_\varsigma}\right]\right]$$

If the relaxation parameter γ is supposed to describe **conversion processes in reversible-adiabatic limiting cases**, its explicit information is necessary for the use of equation (3.s). If not, the unknown $\hat{\gamma}$-value may be determined from the extended L.-P.-Law.
Only in the case of $\hat{\gamma} \equiv 0$ (i.e. no conversions occur: $r \equiv 0 \rightarrow$ all particles are in the ground state) does equation (3.s) change into the classical 'adiabatic law'!

It should be noted that the overdeterminism hidden in classical gas dynamics may, for example, be eliminated by equation (3.s) (see final remarks of this section).

Finally, one should assume a more realistic case in which an **irreversible catalytic process** occurs between state $\underline{\infty}$ and $\underline{F}$. With an isobaric increase of model gas temperature from $T = 0$ K in $\underline{\infty}$ (i.e. $r \rightarrow 0$) to a strongly excited state in $\underline{F}$ with a high temperature, one can use equation (3.q) together with the energy equation of gas dynamics

$$h_\infty + \frac{1}{2} v_\infty^2 = h_F + \frac{1}{2} v_F^2$$

as follows: for an isobaric change of state $v_F \approx 0$ results in $\underline{F}$ from the equation of motion with $v_\infty^2 \ll h_\infty$. From the enthalpy condition $\Delta h = 0$, and using equation (3.q), one can explicitly calculate the 'flame temperature'

(3.t)
$$T_F = \frac{\kappa-1}{\kappa}\left[\frac{1}{1+\gamma_\infty^{-1}} - \frac{1}{1+\gamma_F^{-1}}\right]\theta$$

For calculating the gas dynamic examples it is sufficient to choose a constant for the expression in brackets above. This constant, together with the characteristic vibrational temperature θ and the κ-value for elementary gases with full rotational excitation, gives a combustion temperature T_F with a realistic order of magnitude. Temperatures of approximately 3000 K occur for hydrogen combustion. With $\theta(H_2) = 6322$ K and $\kappa = 1.4$, one arrives at 5/3 for the bracketed value. When this numerical value is used as an approximation for diatomic gases with other κ-values, the practical formula

(3.u)
$$T_F := \frac{5}{3} \frac{\kappa-1}{\kappa} \theta$$

results.

It is obvious that equation (3.u) is actually a **definition** for simply determining T_F. It succeeds in giving a concrete physical basis to a rather abstract thermodynamical process, expressed through the relaxation parameter γ, since it is related to the vibrational degrees of freedom of diatomic gases. Thus it is not justifiable to infer γ_F or γ_∞ using a T_F-value calculated respectively from equation (3.u) or (3.t).

Computation of the conversion parameter $\hat{\gamma}$ with equation (3.s) is possible, assuming known pressure and temperature values in two neighboring states coupled with the constraint $\Delta s = 0$. This possibility reveals the significance of all 'adiabatic laws' (like equation (3.s)) in reversible adiabatic processes with conversion processes.

Finally, an equation for the speed of sound $\underline{a}$ of the gases is required. It is derived from the definition $\underline{a}^2 := (\partial p/\partial \rho)_S$ from equation (3.s). Using the two simple differentials $(\partial p/\partial T)_S$ and $(\partial \rho/\partial T)_S$, the subsequent quotient becomes

(3.v)
$$\underline{a} = \sqrt{\frac{\kappa - \frac{b}{T}}{1 - \frac{b}{T}} RT} \quad ; \quad b := \frac{\kappa - 1}{1-(\hat{\gamma})^{-1}} \theta \quad .$$

In spite of the drastically simplified description of the conversion process, the speed of sound has surprisingly changed compared to the familiar form in classical gas dynamics

(3.w)
$$\underline{a} = \sqrt{\kappa R T} \quad .$$

In chemical relaxation processes, there are comparable values of κ approximately between 1.2 and 1.1. Considering the simplification chosen for the model gas and the substantial improvement in the transparency of the gas dynamics calculations, an approximate value of $\underline{a}$ by equations (3.w) is readily justified. In the case of the J-2 engine and in reference to the states $\underline{F}$ and $\underline{C}$, $\underline{a}$ lies about 8% higher in (3.w) than in (3.v).

In summary, it should be emphasized that the theory presented here offers a means of avoiding the grave errors found in classical gas dynamics. The traditional isentropic flow tube theory (with κ = constant) yields solutions for four properties of state: T, p, ρ and v, yet is based on five equations for continuous distributions[71]. Therefore, the fifth property could be a conversion parameter as $\hat{\gamma}$.

The Earnshaw-paradox (fallen into obscurity), with its solution

$$(3.x) \qquad p = p_0 - \frac{B^2}{\rho}$$

for the adiabatic equation, is a typical result of overdeterminism: no fluid obeys this equation containing the known reference pressure p_0 and the constant mass flow density B. Equation (3.x) directly follows from both the continuity and the Bernoulli equations; other assumptions are not necessary!

APPENDIX 2: **Properties of State of Polynary Fluid Mixtures**

* The fluid state of LH and LOX

In high-performance rocket engines there is an obvious trend toward
ever-higher combustion chamber pressures. In propulsions powered by a
hydrogen-oxygen combustion, however, basic problems arise when feed
pressures extend into the region of hundreds of bars. Unfortunately
none of the current design and calculation fundamentals adequately take
these thermodynamic problems into account. They are treated in this appendix to prevent errors in calculating procedures.

The fluid phase of the hydrogen and oxygen mixture in state **F** at the
combustion chamber entrance has been defined and explained in Chapter 2
of Part I. This is the fluid state reached at the instant both components have been mixed just behind the injection head and before they
have been ignited.

In order to understand the actual relationships involved, the following
table presents the thermodynamic data of the two fuel components used
in the J-2S rocket engine:

Properties of state in F		Oxygen	Hydrogen
temperature	[K]	96.3	150.2
pressure	[bar]	99.1	95.8
density	[mol/l]	35.42	7.43
specific enthalpy	[J/g]	−402.2	−2307.4

The above data, with the exception of the density values, was selected from the JANAF Tables (1975, p.60): they have been converted from US-units to the international metric system. The molar density values were calculated using the precise, material-specific thermal equation of state given by E. Bender (see NEBENDAHL et al. 1987).

This familiar equation of state for systems at rest (see equation (2.8) in Part II) concurs with a number of other precision equations; it rather accurately delineates a broad range of thermal and caloric state relations within a homogeneous substance. Bender's equation of state includes approximately 20 material-specific constants covering individual molecular interactions of every variety – in each state from an ideal gas region through the two-coexistent phases to liquid states and finally to a highly-condensed homogeneous state. The latter state is also called **fluid**.

For a comparison of the JANAF data with values derived from the Bender equation, it is necessary to relate the caloric properties of state (calculated from this equation of state) to the common standard state:

$25 \circ C$; 1 bar; ideal gas state (see PRIGOGINE & DEFAY 1962, pp.122 f).

In the following comparison based on the previously compiled temperatures and pressures, one notes two discrepancies in the molar enthalpies: in the case of O_2 nearly 3%, and in the case of H_2, 14% . At present it is difficult to determine which source of information is more accurate.

Properties of state in F		Oxygen	Hydrogen	Source
standard molar enthalpy	[J/mol]	8680	8467	
standard molar entropy	[J/mol K]	205.0	142.1	
molar enthalpy	[J/mol]	−12870	−4615	JANAF
		−12437	−4050	BENDER
molar entropy	[J/mol K]	96.48	84.82	"
molar mass		32	2	

Whereas the discrepancy for O_2 has been extensively clarified in the highly accurate thermal equation of state formulated by W. Wagner and associates (see NEBENDAHL et al. 1987), the problems with H_2 have yet to be solved.

* Mixture point and thermal equation of state of a real mixture

In order to fix the state $\underline{F}$, it is physically ingenious to assume an **adiabatic-isenthalpic** process for the **mixture** of the two propellant flows. Since the pressures scarcely differ in such a process, the set of equations (1.35) to (1.37) in Part II are reduced in practice to an enthalpy equation for the iterative determination of the mixture's temperature T_F. If kinetic energies relative to the specific enthalpies of the fluid substances are ignored, the iteration formula

(2.a) $\quad \dfrac{O/F}{1+O/F}\, \hat{h}_{LOX}(T_{LOX},\ p_{LOX}) + \dfrac{1}{1+O/F}\, \hat{h}_{LH}(T_{LH},\ p_{LH}) = \hat{h}_F(T_F,\ p_F)$

for calculating temperature T_F of the fluid mixture results from the First Law (see BAEHR 1966, pp.267 f). The mixture pressure p_F for $p_{LOX} \neq p_{LH}$ is iteratively obtained from the entropy balance

(2.b) $\quad \dfrac{O/F}{1+O/F}\, \hat{s}_{LOX}(T_{LOX}, p_{LOX}) + \dfrac{1}{1+O/F}\, \hat{s}_{LH}(T_{LH},\ p_{LH}) = \hat{s}_F(T_F, p_F) + \Delta s_{mix}.$

In the h,s-diagram, mixture point F' lies on the 'mixture line'(see

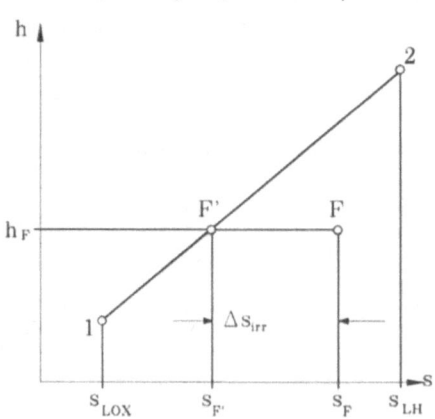

sketch) only if one can asymptotically assume a **reversible mixture.** For the ICP of the rocket engine treated here, this limiting case is fulfilled by definition excluding a **mixing entropy.**[2] The mixture point always lies near the corner point assigned to the larger mass flow rate; with its help, T_F and p_F are fixed.

Utilization of equations (2.a) and (2.b) assumes prior knowledge of the mixture's respective specific enthalpy and entropy. To obtain this information, one can use a theoretical alternative:

(α) the mixture's thermal equation of state $\hat{p}(T, V, n_k)$ as well as the specific heats of the mixture components in the ideal gas limit are known, or

(ß) the excess functions can be modeled.

The relationships between the mixture's equation of state $\hat{p}(T, V, n_k)$ and caloric properties of state can be taken from any good textbook on technical thermodynamics (see PRAUSNITZ 1969, p.40):

* enthalpy of the mixture

$$(2.c) \qquad H = \sum_k n_k \overset{\circ}{h}{}^m_k - \int_\infty^V \left[p - T\left(\frac{\partial p}{\partial T}\right)_{V,n_k} \right] dV + n\, RT(Z - 1)$$

* entropy of the mixture

$$(2.d) \qquad S = \sum_k n_k \overset{\circ}{s}{}^m_k - \int_\infty^V \left[\frac{p}{ZT} - \left(\frac{\partial p}{\partial T}\right)_{V,n_k} \right] dV + R \sum_k n_k \, \ell n\left(\frac{V}{n_k RT}\right) \quad .$$

$\overset{\circ}{h}{}^m_k$ and $\overset{\circ}{s}{}^m_k$ denote the (**ideal gas**) values of the respective molar enthalpy or molar entropy for the k^{-th} component $(k = 1(1)K)$; $Z \equiv pV/n\,RT$ stands for the compressibility factor as a dimensionless expression of the mixture's thermal equation of state. It indicates a problem which has yet to be satisfactorily explained: the interaction between molecules of different components. These interactions are currently described using mixing laws based almost entirely on heuristic arguments (see LUCAS 1986, pp.221 f).

For ideal gases $(Z \equiv 1)$ as mixture components, the two integrals given above are equal to zero. If both equations (2.c) and (2.d) are divided by the total mole number $\sum n_k$ for this limiting case, they reduce to equations (3.60) and (3.61) of Part II.

* Ideal mixtures and excess functions

For the distinct **classification** of non-ideal mixtures, two special de-
scriptive forms are often more suitable than a merely general reference
to the mixture's respective thermal equation of state. Definitions of
the

      (1) **ideal mixture**

      (2) **excess functions**

are needed: both are based on the concrete modeling of the molar chemical potential μ_k^m of the mixture's k^{-th} component.

The ideal mixture (index id) consists of components which all have a
molar chemical potential in the form

$$(2.e) \qquad \mu_{k,id}^m := \hat{\mu}_k^{\dagger m}(p, T) + R T \ln x_k \quad ; \quad k = 1(1)K \quad .$$

Characteristic for these potentials is the additive decomposition of
the dimensionless expression $\mu_{k,id}^m/R T$ into one term dependent only on
T and p, and another logarithmic term which is dependent only on the
mole fraction x_k of the respective k^{-th} component.

This definition leads to the following expressions for the thermodynamic functions of an **ideal mixture**:

□ molar free enthalpy

$$(2.f) \qquad g_{id}^m = \sum_k x_k \mu_k^{\dagger m} + R T \sum_k x_k \ln x_k$$

□ molar enthalpy

$$(2.g) \qquad h_{id}^m = \sum_k x_k h_k^{\dagger m}$$

□ molar entropy

$$(2.h) \qquad s_{id}^m = \sum_k x_k s_k^{\dagger m} - R \sum_k x_k \ln x_k$$

□ molar volume

$$(2.i) \qquad v_{id}^m = \sum_k x_k v_k^{\dagger m} \quad .$$

It is important to stress that the properties of state marked with the index † by no means refer to ideal gas states, but rather to the **real** state of the respective fluid component. Equations (3.60) and (3.61)[73] are special cases of equations (2.g) and (2.h) for ideal gases, i.e. for real gas components whose molecular interactions between each component's own and other particles can be ignored.

Equation (2.f) also indicates that the chemical potential $\mu_{k,id}^m$ is identical with the molar free enthalpy $g_{k,id}^m$ of the k^{-th} component defined by equation (2.e).

Few mixtures are near ideal in a fluid state. Real mixtures are characterized by respective formulas for the molar chemical potentials of each of its components: the expressions are formulated analog to equation (2.e) (see ROWLINSON 1959, p.124):

$$(2.j) \quad \mu_k^m - \hat{\mu}_k^{\dagger m}(p, T) := R T \ln(x_k\gamma_k) := R T \ln a_k \; ; \quad k = 1(1)K.$$

This equation simultaneously defines the **activity** a_k and the **activity coefficient** γ_k of the k^{-th} component; these properties contain all those influences of intermolecular forces causing deviations from the ideal mixture.

These deviations are described through **excess functions** (index E): individually they are defined for a real mixture's most important caloric properties of state as follows:

$$(2.k) \qquad \frac{h^{m,E}}{R\,T} = - \sum_k x_k \left(\frac{\partial \ln \gamma_k}{\partial \ln T}\right)_{p,x_k}$$

$$(2.l) \qquad \frac{s^{m,E}}{R} = - \sum_k x_k \left(\frac{\partial}{\partial T}(T \ln \gamma_k)\right)_{p,x_k}$$

$$(2.m) \qquad v^{m,E} = R\,T \sum_k x_k \left(\frac{\partial \ln \gamma_k}{\partial p}\right)_{T,x_k}$$

A real mixture's specific limiting case is of particular significance for current rocket theory. This case is termed **athermal mixture** and is defined by

(2.n) $$\left| h^{m,E} \right| < T \left| s^{m,E} \right| \quad .$$

Deviations from the ideal state are caused by entropy effects. Equation (2.n) implies that both the molar 'mixing heat' $h^{m,E}$ and the molar excess volume $v^{m,E}$ may be ignored.

The constraint (2.n) is prompted by experience: it gives a large number of mixtures that can be classified as athermal in a wide range of variables (see PRIGOGINE & DEFAY 1962, pp.421 f). Pertinent experimental information is still, however, rather meagre. Due to this gap, theoretical comprehension - such as the explanation of excess properties as a result of intermolecular forces (see LUCAS 1986, p.420) - of such fluid mixtures is still limited. This is particularly true in the case of the LH-LOX mixture vital for rocket engine technology. Since it is extremely dangerous to experiment with this mixture, there is not a single test data available offering precise information on its thermal and caloric behavior in a fluid state. Tendencies may, at most, be inferred from analog examples. Naturally such inferences do not offer reliable equations of state for the LH-LOX mixture. Application of mixing laws to the indeterminate coefficients of an appropriate thermal equation of state at least allows approximations of the excess properties. With these approximations, in principle one can at least estimate pressure and temperature in state $\underline{F}$; the values lie, depending on the mass flow ratio, between the respective pressure and temperature values of the unmixed fluid components immediately behind the injection head.

* The fluid state $\underline{F}$

In the following table the left sides of both equations (2.a) and (2.b) are evaluated for the specific enthalpy h_F and specific entropy s_F of the LH-LOX mixture, based on a negligible irreversible entropy increase. The pressure and temperature data (with respective caloric data) used here are based on the data occurring in the J-2 rocket engine injection plane: the value O/F = 5.5516 was chosen as the mass flow ratio. The enthalpy data is from NASA-papers; density and specific entropy of the LOX and LH were calculated from the respective thermal equations of state by W. Wagner and E. Bender (see NEBENDAHL et al. 1987).

Properties of state in $\underline{F}$		LOX	LH	LH-LOX-Mix
molar mass	kg/kmol	32	2	9.73
mole fraction	–	0.2576	0.7424	–
pressure	bar	←———	53.3	———→
temperature	K	90.18	20.17	?
density	kg/m³	1153	76.71	988.7
specific enthalpy	kJ/kg	−406.1	4512	1033
specific entropy	kJ/kg·K	2.92	16.98	5.55
flow velocity	m/s	–	–	1.38

In the mixture's specific entropy the unavoidable share $s^{mix} := (R/\psi) \Sigma \chi_k \ln \chi_k$ was incorporated respective to the mole fractions χ_k (see equation (2.h); in the example, this was equal to -0.4876 kJ/kg·K).

The flow velocity in $\underline{F}$ is obtained through the equation for the steady mass flow rate and the data in Table 2 of Part I. The low velocity justifies an approximated coupling

$$(2.o) \qquad h_F = h_C + \tfrac{1}{2} v_C^2$$

of the specific enthalpies in states $\underline{F}$ and $\underline{C}$. Since at least h_F and p_F offer correct information on the mixture (although the mixture temperature T_F cannot be calculated with equation (2.a) due to the lacking mixture equation of state!), one arrives at a **first approximation** T_{ad} for the temperature of the ignited mixture in $\underline{C}$, assuming $\tfrac{1}{2}v_C^2 \ll h_C$. This calculation is based on the AFC method: it produces precise T_C values at cross section ratios of $\alpha \geq 2.5$, along with the respective equilibrium mole and mass fractions. With parameters $\alpha < 2.5$, the AFC method gives a temperature $T_{ad} > T_C$. If one determines the respective specific entropy $\hat{s}_{ad}(T_{ad}, p_F)$ of the hot gas mixture with p_F (and the mass fractions $\hat{\omega}_{ad}(T_{ad}, p_F)$ using equation (1.30-II) and (1.31-II), one obtains a numerical value of approximately 18.65 kJ/kg·K for the J-2 rocket engine.

The conflict is obvious. In an adiabatic-isenthalpic combustion, and based on equation (2.b), the relationship

$$(2.p) \qquad \hat{s}_F(T_F, p_F) + \Delta s_{mix}^{irr} = s_C \approx s_{ad}$$

apparently gives the correct expression for the isentropic condition between the fluid state $\underline{F}$ and the gas state in (for example) $\underline{C}$. Calculating the specific entropy s_F in greater detail with

$$(2.q) \qquad \hat{s}_F(T_F, p_F) = \sum_k s_k \omega_k = \hat{s}_{id}(T_F, p_F) + \hat{s}^E(T_F, p_F) \quad ,$$

then the following holds: at a negligeable specific excess entropy s^E and an insignificant entropy increase Δs_{mix}^{irr} resulting from irreversible mixing, ignition, diffusion, turbulence etc.[74], the absolute value 5.55 kJ/kg·K of $\hat{s}(p_F, O/F, T_{LOX}, T_{LH})$ given above approaches the value $\hat{s}_{id}(T_F, p_F)$ of an ideal mixture.

Since one lacks reliable information on both s^E and Δs_{mix}^{irr} , this premise cannot be verified. The concept of the ideal comparative process allows one to disregard the irreversible entropy increase. If this is done, however, one **apparently must expect an unusually high specific excess entropy s^E of the fluid mixture in $\underline{F}$** . In the example above, the value of s^E lies at approximately 18.65-5.55 ≈ 13 kJ/kg·K, provided the approximation $\hat{s}_{id}(T_F, p_F) \approx \omega_{LOX}\hat{s}_{LOX}(T_{LOX}, p_F) + \omega_{LH}\hat{s}_{LH}(T_{LH}, p_F) - s^{mix}$ is acceptable.

In this expression the extreme **reactive affinity of hydrogen** is taken into account

The process described here can be summarized by the reversible limit of equation (2.p)

$$(2.r) \qquad \boxed{\hat{s}_F(T_F, p_F) \overset{!}{=} s_C} \qquad :$$

the left side of this constraint contains a property which, as previously shown, is not directly available at present in the case of a hydrogen-oxygen **fluid** mixture. One is thus forced to get the lacking information indirectly through calculation of the gas mixture's specific entropy s_C in state $\underline{C}$.

This calculation is done according to the scheme — equation (3.18) — explained in Section 3.2 of Part II. Specific entropy $\hat{s}(T_{ad}, p_F)$ is used as the **initial value** of the iteration; this value is established by the adiabatic temperature T_{ad} derived from equation (2.o) (under the premise $v_C = 0$ for velocity), by the pressure p_F and the respective equilibrium concentrations $x_{k,ad}(T_{ad}, p_F)$. Only in the limiting case $\alpha := A_C/A_T \to \infty$ is $s_{ad} = s_C$ asymptotically valid. In the iteration scheme (3.18-II), constraint

$$(2.s) \qquad \hat{s}(T, p, x_k) - \hat{s}_{ad}(T_{ad}, p_F, x_{k,ad}) = 0$$

and ten other relationships are used to calculate the coordinates T_C, p_C and $x_{k,C}$ and three Lagrange multipliers.

The resulting pressure p_C is, respective to the given area ratio α, considerably lower than the pressure p_F. Together with the temperature T_C and the established parameter value for the steady mass flow rate $\dot{m}$, one can determine the velocity $v = \sqrt{b}\,\rho^{-1}$. Here the mass density ρ in $\underline{C}$ is known through the thermal equation of state. This flow velocity yields the specific kinetic energy $\frac{1}{2}v_C^2$. If one subtracts this energy share from the fluid mixture's given specific fluid enthalpy h_F, one can calculate (using this reduced specific enthalpy and pressures p_F and p_C) a further approximation (considering the finite chamber cross section) for the unknown specific entropy s_F, using the method recommended in Section 3.6. The iteration procedure in Chapter 3 of Part II is initiated with this approximation for the fluid entropy s_F.

It is evident that **the value s_{ad} is merely provisional: under no circumstances may it be considered generally equal to the specific entropy s_F of the fluid mixture!**

The calculation procedure is sufficient as long as one assumes (in agreement with the experimental findings from engine tests) that the gaseous polynary mixture along the combustion chamber-nozzle configuration behaves like an ideal gas. This assumption certainly can no longer be considered valid when the combustion chamber pressure rises above 500 bar.

Notes

1   In all six rocket engines considered in this study, one finds de-
    viations of less than 0.5% from the mean value $c^* = 2327$ m/s; only
    the HM7-B engine deviates barely 1%! (See Table 2).

2   Naturally there is a direct relationship between definitions (1.1)
    and (1.6) if one agrees on an effective exhaust velocity $v_E^o := v_E +$
    $(p_E - p_U) A_E / \dot{m}$ for the formula $S_E := \dot{m} v_E^o$.

3   Further sources of energy loss, such as the boosting of the fuel
    pressure from tank to combustion chamber level, should not be as-
    sessed with the comparative process; the chosen system definitions
    have to be modified.

4   Cooling techniques of the SSME are technologically based on the
    MBB engine CS50K-H, a design in which 400 bar combustion chamber
    pressure was reached with a LH-LOX combustion!

5   Confirmation of Prozan's prognosis – though on a completely dif-
    ferent physical basis – is one of the remarkable results of the
    study (see Section 4.3 in Part II).

6   Modification of this state $\underline{F}$ for conditions in which $\frac{1}{2}v^2 \geq h$ is
    valid is a decisive prerequisite for realistic designs of new
    flight systems such as the SÄNGER type. A highly simplified but
    theoretically relevant model for an air-augmented rocket was ear-
    lier proposed by RICKEARD (1973,pp.19 f). The air is diffused to a
    specified Mach number; free stream values up to 2.0 are discussed.

7   Prozan's postulate "that the throat choking condition acts as a
    constraint to the combustion process" (PROZAN 1982, p.4) apparent-
    ly expresses the same state of facts (see Section 1.3 in Part II).

8   Prozan comments on the 'comparatively simple equilibrium equa-
    tions' as follows: "The equilibrium methodology is well accepted
    but, because of its importance, a reexamination of the concepts is
    justifiable." (PROZAN 1982, p.3).

9   The conditions of adiabatics can be abandoned under certain situa-
    tions (see Section 3.7 in Part II).

10 The agreement corresponds to the "staged combustion cycle" of the
    SSME; it can be modified for a "gas generator cycle with dump
    cooling of nozzle", as is done for the HM 60.

11 This presumption or its extension, the demand for reversibility,
    implies a steady state: reversible processes must always be
    steady, yet steady processes are normally irreversible. In the
    framework of **linear** irreversible thermodynamics they are typified
    by a minimum of the local entropy production density.

12 For this study pressure p means the thermodynamic pressure – that
    is, through a thermal or caloric equation of state coupled with
    other properties of state. Expressions such as static, dynamic,
    total pressure, etc., are not used.

13 Although K.N.C. Bray emphasized the eigenvalue-character of $\dot{m}$ in 1970 (Bray in WEGENER 1970, p.82), this criterium was ignored even in later editions of the Lewis codes. This example of negligence shows how necessary continuous updating and controls of such important programs are: only with such systematic checks can the current relevance and reliability of influential planning concepts be guaranteed!

14 The postulate $p_F \geq p_U$ is sufficient to eliminate entropy production (resulting from compression shocks: see BARRèRE et al. 1960, pp. 76 f) just ahead of the nozzle exit section.

15 Without sacrificing accuracy in the calculation of T_F, the usual 'flame-state' F may be characterized by the infinite cross section area $A \rightarrow \infty$ and thus by the pertinent limiting value $v \rightarrow 0$.
   For the determination of non-zero velocity values in F, however, a finite cross section area is assigned in order to formally fulfill the demand $\dot{m}$ = constant. The difference is relevant only for the pressure drop; for realistic calculations only the 'fluid state' F is important (see Part II and Appendix 2).

16 Equation (3.13) is a significant prerequisite for the following equations in this section. Equivalent relationships for $\dot{m}$ and S_E can also be derived without considering equation (3.13).
   ADAMS (1987) followed this alternative path in a separate study. The concept presented here is based on equation (3.13) with reference to the $_\wedge$investigation of Prozan (see Part II), in which this relationship $\alpha(M_C, \kappa)$ plays an important role.

17 "ERROR 1" can easily be expressed (with the information from Appendix 1) as the respective difference in entropy.

18 Compare with equation (3.s) in Appendix 1.

19 Equation (3.18) is a typical result of a 'simple fluid' view. By renunciating of a realistic multicomponent mixture theory, the important influence of the fuel-oxidizer mass flow ratio on nozzle temperature distributions is eliminated (see Sec. 4.3 in Part II).

20 In gas dynamics the famous Bernoulli-equation results from the classic isentropic relation (3.7) and the energy equation (3.5) (see ZIEREP 1976, p.51).

21 The thrust in unit kg is obtained with Ω_S/g_0 (see equation 1.3).

22 The remainder of the quotation can be found in Section 3.2.

23 See the bibliography in ÜÇER et al. 1985, p.824

24 The process of non-equilibrium flow with dissociating gases can be better understood with the help of the Lighthill model (see Bray in WEGENER (Ed.) 1970, pp.77 f).

25 "Constraint Entropy Maximization Concept Workshop Meeting to Assess its Validity, held February 27 and 28, 1985, at Continuum Inc., Huntsville, Ala."
   At the author's request, Prof. R. Waibel of the University of Federal Armed Forces Munich also participated in the plenar meetings.

26 'Reaction equations', as one knows, are not equations in the usual mathematical sense: they classify individual components whose names are symbolized with $\mathfrak{C}_k$ (for example, H_2O as steam), and separated into reactants ($\nu_{kr} \leq^k 0$) and (reaction) products ($\nu_{kr} \geq 0$).

27 In contrast to equation (3.10) in Part I, the thermal equation of state now contains the universal gas constant $\mathbf{R}$ coupled with the specific gas constant R with $R = \bar{R}/\psi$ over the mean molar mass ψ.

28 Both authors also reported on other algorithm solutions for chemical equilibriums.

29 An example for the HM 60/1 engine: $h^m(LH) = -8348.5$ J/mol at 34.5 K, $h^m(LOX) = -12455.9$ J/mol at 95.5 K; the pressure is 103.6 bar each; $\dot{m}(LOX)/\dot{m}(LH) = 5.89$ gO/gH . With $\psi(H) = 1$, $\psi(O) = 16$ and ψ according to equation (1.23.2), the negative value -938.5 J/g results as the specific enthalpy $h = h^m/\psi$ in state $\underline{F}$; this result corresponds to published values.

30 The partial differentiation denotes constancy of all ℓ-marked variables of $\mathcal{L}_s(\mathbf{n})$ other than those with an index such as $k \neq \ell$ or $e \neq \ell$ with respect to which the differentiation is being carried out. The first equation of system (1.41) was corrected in the λ^E_M and λ^M terms, compared to the original work (PROZAN 1969, p.15)!

31 The Fundamental Relations are basic elements of thermodynamics (see CALLEN 1966, pp.25, 31 f, 36).

32 Some authorities offer notable examples of expensive misunderstandings with statements regarding the amount of 'heat' in the energy equation: "In combustion, for instance, it would correspond to the heat supplied by the chemical reaction." (BARRèRE et al. 1960, p.84). Such an interpretation is wrong and prevents the concrete application of an adequate mixture theory for conversion processes.

33 The data has been graciously supplied by MBB Co., Munich.

34 The polynomial representations for the partial molar enthalpy and entropy of the components used in the NASA-Lewis Code have not been checked by the editors in over twenty years, let alone altered.

35 In this sequence $\Delta\epsilon_{j=0} = 0$ is valid per definition. For constant values $\Delta\epsilon$ the sequence $\epsilon_{j+1} := \epsilon_j + j \Delta\epsilon$ is used.

36 The information also was supplied by MBB Co., Munich.

37 H. Tetens (in MENNE (Ed.) 1987, p.176 ff) offers an informative contemporary analysis of the debate on the reality of micro-objects, primarily from the viewpoint of P.W. Bridgman's "operationalism" and H. Dingler's scientific-theoretical term of reality. It is an excellent supplement to d'Espagnat's profound essay on quantum theory.

38 H. Dingler provided a sober analysis of the ideological foundations of similar principles before WW II (see DINGLER 1939, pp.4-8 and 25-26).

39 Although time t does not appear explicitly in the fundamental thermodynamic dependencies, it is factually not merely a 'passive' parameter: the nature of the space of state itself is dependent on t through the time scale of the observer. Fundamental concepts such as adiabatics, closure, process, equilibrium etc., all involve comparisons between various scales for relaxation processes (see WOODS 1975, p.70). On the micro-level, the possibility of describing the exchanges and transitions occurring in various relaxation times within the framework of a macro-process is founded on a fundamental second (together with Heisenberg's) principle of uncertainty, as a consistency condition (see STRAUB 1987, pp.216-7). The research of L.C. Woods, in particular, demonstrates the problems of the approved 'principle of material objectivity' in Rational Thermodynamics (see DELGADO DOMINGOS et al. (Eds) 1974, p.38).

40 Contrasting sharply to this approach there is still a number of strictly-limited definitions of what should be considered 'thermodynamics' (see SILHAVY and FEINBERG & LAVINE in SERRIN (Ed.) 1986, pp.35 and 52 f).

41 The point marks the scalar product of both vectors **v** and d**P**.

42 The term 'extensive' is used in a more general sense here than usual: the material nature of a property is a sufficient yet not necessary criterium for it being extensive (see FALK & RUPPEL 1976, p.88).

43 For example, the entropy in reversible processes, or also the hypercharge and the isospin in elementary particle processes.

44 The physical substance of the GFE $dE = \mathbf{v} \cdot d\mathbf{P} - \mathbf{F} \cdot d\mathbf{r}$ has traditionally dominated classical theoretical mechanics, where it is denoted as **Hamiltonian system** concerning all mass point-field-systems.

45 The useful conversion formulas can be used (μ_k^t and μ_k^m, respectively: chemical potential per particle and per mole, respectively):

$$m_B^{-1}\sum_k \mu_k^t N_k = m_B^{-1}\sum_k N_L \mu_k^t N_k/N_L = \sum_k \mu_k^m n_k/(n\psi) = \sum_k (\mu_k^m/\psi_k)x_k\psi_k/\psi = \sum_k \mu_k \omega_k \quad .$$

46 The modern version of the old situation can be described as follows: the expression for the pressure tensor in the Navier-Stokes motion equation is no longer physically 'derived', but rather 'defined'. The theoretical reasoning is achieved 'structurally', the invariance theory of tensors traditionally 'explains' the appearance of viscosity coefficients in the pressure tensor. As far as the actual contents are concerned, there is nothing which Stokes wouldn't have known!

47 L. BOLTZMANN & J. NABL: Kinetische Theory der Materie; Encyklop. d. math. Wissensch. Bd. V$_1$, Heft 4 (1907), p.545.

48 The thermal equations of state for real fluids and their mixtures are experimentally accessible only for the state of rest (see equation (2.8)). In the proximity of critical points, however, the energy form **F**·d**r** gains influence according to the GFE (2.5) in spite of the constant gravitational field $F \equiv g(\mathbf{r})$, since density

fluctuations occur throughout the full height of the test chamber; in fact, $p(T, \rho)$ results. In space experiments, one can determine the 'pure' equation of state $p(T, \rho)$ with the constraint $g(\mathbf{r}) \to 0$ near such critical points as well.

49 Its definition allows additional possibilities:

$$i \cdot \left[\frac{\partial \phi}{\partial s}\right]_{\rho, \omega_k} \lessdot T \quad ; \quad \phi \cdot i = \phi \cdot (v - \phi) \approx \phi \cdot v \quad ; \quad \partial_t \phi \equiv 0$$

50 In a personal remark (1985).

51 TRUESDELL (1984, pp.426-427), for example, also agrees: "it is a postulate, an assumption, the definition of a model."

52 Volume of **one** mole $V^m := 22.414 \text{ dm}^3/\text{mol}$ refers to an ideal gas state at 0°C and 101325 Pa; Avogadro number $N_A := 6.02204 \cdot 10^{23}$ particles per mole.

53 According to kinetic gas theory, the factor f lies on the order of one; the value $f \equiv 2$ is chosen as an average between 5/3 and 5/2.

54 With minor differences in the values for the parameters κ and ν_M.

55 Prozan gave a similar expression for the auxiliary property ϕ_k^E, yet with typographical and mathematical errors (PROZAN 1982, p.7).

56 Since the sign T also stands for a temperature derivative in the following text, every property which is used for the state $\underline{I}$ or another state will be indexed with this sign $\underline{I}$ or the sign of the respective state. This arrangement is valid only for the remainder of this chapter.

57 For compressible fluids affected by friction, the speed of sound $v_{\underline{I}}$ is primarily dependent on the process realization.

58 The notation is commented in the list of symbols.

59 The nozzle exit pressure can also be a functional of O/F instead of p_F.

60 Inquiries should be addressed to the author.

61 Co-author: R. Waibel

62 For example, with regard to the choice of the maximum value of the auxiliary function θ_1!

63 If the given value of ß is smaller than one according to equation (3.77), one must assume that a part of the diverted energy is not regeneratively returned, but rather lost to the surroundings.

64 An example: with a hydrogen initial temperature of 40 K and an exit temperature of 100 K, the value of the enthalpy difference is $\Delta h_c = 901 \text{ kJ/kg}$ in an isobaric flow process (see MIELKE 1986, p.213).

65 Efficiencies often become higher than unity concerning the power of rocket propulsions (see BARRèRE et al. 1960, p.31)! They are

apparently reached through the 'freely available' mass flow rates; in the SSME the 'increase' reaches 115% (RUPPE 1982, p.74)!

66 With more realistic values such as ß = 0.1 (or θ = 0.01, respectively) for the cooling coefficient, one obtains I = 461.7 (or I = 460.9, respectively) as the specific (vacuum) impulse. These values confirm the above-mentioned linear formulation (4.3.1).

67 Co-author: V. Lippig

68 Lunar Modul Ascent Engine

69 In a by-pass flow procedure, in contrast to a main flow procedure (used in the SSME, for example), the mass flow rate is slightly increased over the 'eigenvalue' $\overset{\bullet}{m}$ in the expansion section of the Laval nozzle (examples: HM 60/1: $\Delta\overset{\bullet}{m}$=1.8 kg/s; J-2S: $\Delta\overset{\bullet}{m}$=5.1 kg/s; see Table 2 in Part I).

70 In the development of the space shuttle orbiter, the share of the main engine costs was 50% (1982: 1.4 billion US $).

71 This statement is not valid for normal shock investigations. The Rankine-Hugoniot relations are derived from steady, one-dimensional **viscous** conservation laws assuming vanishing local derivatives for the infinite limit of both the positive and negative directions. Consequently, these relations are consistent with irreversible entropy changes across shocks.

72 In real mixture processes, the entropy increase $\Delta s(mix) \geq 0$ expresses omnipotent irreversibilities; the mixture point F then lies right of F' at the level of its specific enthalpy h_F.

73 In equations (3.60) and (3.61) index ° marking has not been used, since there is no danger of confusion as in equations (2.c) and (2.d).

74 Compared to the expression $\Delta s(mix)$ used in equation (2.b), the term $\Delta s(irr/mix)$ no longer contains the specific mixing entropy $\Delta s(mix)$; this is subtracted from the sum $s_{LH}(T_{LH}, p_F) + s_{LOX}(T_{LOX}, p_F)$ (see equation 2.h).

INDEX

* Index of Names